CELL BIOLOGY RESEARCH PROGRESS

BETA-GALACTOSIDASE

PROPERTIES, STRUCTURE AND FUNCTIONS

CELL BIOLOGY RESEARCH PROGRESS

Additional books and e-books in this series can be found on Nova's website under the Series tab.

CELL BIOLOGY RESEARCH PROGRESS

BETA-GALACTOSIDASE

PROPERTIES, STRUCTURE AND FUNCTIONS

ELOY KRAS
EDITOR

Copyright © 2019 by Nova Science Publishers, Inc.

All rights reserved. No part of this book may be reproduced, stored in a retrieval system or transmitted in any form or by any means: electronic, electrostatic, magnetic, tape, mechanical photocopying, recording or otherwise without the written permission of the Publisher.

We have partnered with Copyright Clearance Center to make it easy for you to obtain permissions to reuse content from this publication. Simply navigate to this publication's page on Nova's website and locate the "Get Permission" button below the title description. This button is linked directly to the title's permission page on copyright.com. Alternatively, you can visit copyright.com and search by title, ISBN, or ISSN.

For further questions about using the service on copyright.com, please contact:
Copyright Clearance Center
Phone: +1-(978) 750-8400   Fax: +1-(978) 750-4470   E-mail: info@copyright.com.

NOTICE TO THE READER

The Publisher has taken reasonable care in the preparation of this book, but makes no expressed or implied warranty of any kind and assumes no responsibility for any errors or omissions. No liability is assumed for incidental or consequential damages in connection with or arising out of information contained in this book. The Publisher shall not be liable for any special, consequential, or exemplary damages resulting, in whole or in part, from the readers' use of, or reliance upon, this material. Any parts of this book based on government reports are so indicated and copyright is claimed for those parts to the extent applicable to compilations of such works.

Independent verification should be sought for any data, advice or recommendations contained in this book. In addition, no responsibility is assumed by the Publisher for any injury and/or damage to persons or property arising from any methods, products, instructions, ideas or otherwise contained in this publication.

This publication is designed to provide accurate and authoritative information with regard to the subject matter covered herein. It is sold with the clear understanding that the Publisher is not engaged in rendering legal or any other professional services. If legal or any other expert assistance is required, the services of a competent person should be sought. FROM A DECLARATION OF PARTICIPANTS JOINTLY ADOPTED BY A COMMITTEE OF THE AMERICAN BAR ASSOCIATION AND A COMMITTEE OF PUBLISHERS.

Additional color graphics may be available in the e-book version of this book.

Library of Congress Cataloging-in-Publication Data

ISBN: 978-1-53615-605-8

Published by Nova Science Publishers, Inc. † New York

CONTENTS

Preface		**vii**
Chapter 1	Microbial Beta-Galactosidase: Production, Bioprocess Parameters and Downstream *Tatiane A. Gomes, Michele R. Spier and Michelle R. F. Vaz*	**1**
Chapter 2	Survey of β-galactosidases Properties: Applications to Transglycosylation Process *Cecilia Porciúncula González, Cecilia Giacomini and Gabriela Irazoqui*	**65**
Chapter 3	Galacto-Oligosaccharide Synthesis by Transgalactosylation Activity of β-Galactosidase: Recent Trends, Challenges and Future Perspectives *Milica Simović, Marija Ćorović, Dejan Bezbradica, Ana Milivojević and Katarina Banjanac*	**117**

Chapter 4	Screening and Identification of β-Galactosidase Producing Microorganisms from Yak Yoghourt in China's Gannan Pasture *Weibing Zhang, Yingying Cao, Kaiyong Wen, Lei Cao, Jiang Ma and Qiaoqiao Luo*	**167**
Index		**185**
Related Nova Publications		**191**

PREFACE

β-galactosidase is an enzyme responsible for catalyzing the hydrolysis of the lactose β-1,4 linkage into α-D-glucose and β-D-galactose.

β-galactosidase is found in plants, animals and microorganisms. In *Beta-Galactosidase: Properties, Structure and Functions*, the authors discuss the main microorganisms that produce β-galactosidase, the characteristics of the culture media, bioprocessing parameters, the most relevant downstream steps used in the recovery of microbial β-galactosidase, as well as the main immobilization techniques.

Next, this compilation examines β-galactosidases classification, mechanisms, characterization and applications.

Recent advances in galacto-oligosaccharides production and purification are also covered, and different challenges and future perspectives are discussed.

In the closing study, strains with β-Galactosidase activity were isolated from 25 Yak Yoghourt samples collected from the Gannan pasturing area of Gansu Province. An efficient β-galactosidase producing strain SYA2 screened out from 21 strains was identified as Enterobacter sp. by means of morphological feature observation, physiological and biochemical characteristics measurement, and 16S rDNA sequence analysis.

Chapter 1 - β-galactosidase (β-Gal; β-D-galactoside galactohydrolase; EC 3.2.1.23, popularly known as lactase) is an enzyme responsible for catalyzing the hydrolysis of the lactose β-1,4 linkage into α-D-glucose and β-D-galactose. β-galactosidase is found in plants, animals and microorganisms. The immobilized form is widely used in the food industry to reduce the lactose content of dairy products, prevent problems associated with lactose crystallization, and increase sweetness, flavor and solubility. Only microbial β-galactosidase is used for industrial purposes due to its viability and lower cost when compared to other sources. Yeast, filamentous fungi and bacteria are potential producers of intra or extracellular β-galactosidase. Intracellular β-galactosidase is obtained by the permeabilization or disruption of the cell wall membrane after the application of mechanical or non-mechanical methods such as sonication and solvents, respectively. The enzymatic production by microorganisms occurs through submerged fermentation (SmF), in which complex culture media, which are rich in lactose (carbon source) and amino acids (provided by nitrogen sources), are used in the production process; it is necessary to control parameters such as pH and temperature. Both the temperature and optimum pH used in enzymatic production vary according to the microorganism. The same occurs for processes which may, or may not, use agitation and aeration because each microorganism has different conditions of production and oxygen tolerance. In addition to lactose, other components are essential for the catalysis of β-galactosidase, such as metal ions (potassium, manganese, and magnesium), which act as cofactors in enzymatic activity, while glucose, galactose and calcium are known as competitive inhibitors of β-galactosidase. Higher levels of β-galactosidase production are achieved when parameters such as the maximum specific microbial growth rate and yield factors are evaluated and controlled. In addition to the production process, the separation and recovery of enzymes must be controlled due to the high proportion of water in the culture medium and the presence of organic and inorganic molecules, as well as extracellular and intracellular metabolites. The purification process is the most expensive stage of enzyme production, accounting for approximately 80% of the total cost. Downstream can be divided into four general steps: clarification (separation of cells and fragments from the culture medium); low-resolution and high-resolution purifications; and the finishing and packaging of the product. The choice of the purification steps will depend on the application of the biomolecule. For industrial applications, immobilized enzymes are preferable due to their high levels of stability, easy separation from the reaction mixture, and reusability. This chapter discusses the main microorganisms that produce β-galactosidase, the characteristics of the culture media, bioprocessing parameters, and the most relevant downstream steps used in the recovery of microbial β-galactosidase, as well as the main immobilization techniques that are used.

Chapter 2 - β-galactosidases (β-D-galactohydrolase, EC 3.2.1.23) are enzymes that hydrolyzes terminal β 1-4 galactosides. They belong to the GH 1, GH 2, GH 35 and GH 42 of the GH-A superfamily of glycoside hydrolases. They are widely distributed in nature and can be find in animals, plants and several microorganisms (yeast, fungi and bacteria). In nature, they function as hydrolases. In plants, they remove terminal β 1-4 galactose from polymers containing galactose, in animals and microorganisms they catalyzes the hydrolysis of lactose in galactose and glucose. The most studied β-galactosidases are those from microorganism, such as *Escherichia coli*, *Aspergillus oryzae*, *Bacillus circulans*, *Streptococcus thermophilus*, *Kluyveromyces lactis*. This allows a collection of β-galactosidases with different properties (as optimum pH and temperature, thermal stability, stability against pH, denaturing agents, etc.), which offer a great versatility of conditions to use these enzymes for different applications. One of the first biotechnological application of these enzymes was for lactose hydrolysis in order to produce lactose reduced milk and dairy products for consumption of lactose intolerant people. Besides, a reduced lactose content in dairy product avoids lactose crystallization improving their organoleptic properties. Nevertheless, under adequate conditions they are able to catalyze the synthesis of oligosaccharides and galactosides. Two methods have been used for this purpose, reverse synthesis and transglycosylation. The first one involves the inversion of the hydrolysis reaction starting from a mixture of monosaccharides and nucleophilic hydroxylated molecules. This method requires high concentration of reactants as well as the presence of co-solvents in order to reduce water activity. On the other hand transglycosylation mechanisms is kinetically controlled and catalyzes the transfer of a galactosyl moiety from a galactosyl donor (lactose, *ortho*-nitrophenyl-β-D-galactopyranosyde, lactulose) to an hydroxylated nucleophile. Aliphatic alcohols, hydroxylated aminoacids such as serine and threonine, polyols, flavonoids, and monosaccharides among others have proved to be good glycosyl acceptors. The use of the transgalactosylation mechanism is an excellent alternative to the complex chemical synthesis as it allows the synthesis of β-galactosides and β-galactooligosaccharides in a single step preserving the anomeric center configuration. Even though glycosidases are stereospecific they are not always regiospecific and so the generation of isomers mixtures could take place, depending on the structure of the acceptor molecule as well as on the source of the β-galactosidase. Regarding synthetic applications, they have been used for enzymatic synthesis of: galactooligosaccharides with prebiotic properties as food additives; galactooligosaccharides as building blocks for synthesis of branched oligosaccharides; galactosides with potential activity as galectin inhibitors or antitumor agent; alkyl galactosides as non-ionic surfactants. β-galactosidases have also been used for galactosylation of drugs in order to improve their hydrophilicity and bioavailability. In this chapter the authors will make an update regarding β-galactosidases classification, mechanisms, characterization of their properties and applications.

Chapter 3 - The growing interest in functional foods and their beneficial effects on human health, yielded due interest in the field of obtaining novel bioactive ingredients featuring enhanced physiological and physical/chemical characteristics. Accordingly, the considerable attention of the scientific community during the last decades was attributed to *β*-galactosidase synthetic activity. Namely, under the specific conditions, *β*-galactosidase catalyzes reaction of transgalactosylation resulting in the formation of a diverse mixture of highly valuable products named galacto-oligosaccharides (GOS). Owing to the specific structure, GOS are primarily recognized as products with the pronounced prebiotic activity and favorable impact on overall human well-being, however, their wide application in the food and pharmaceutical industry is additionally ensured by their high thermal and acid stability, excellent taste quality, low sweetness and caloric value. Throughout this chapter, different aspects related to the recent advances in GOS production and purification, as well as their application with emphasis on their prebiotic role, will be covered. Likewise, different challenges and future perspectives will be discussed.

Chapter 4 - Strains with β-Galactosidase activity were isolated from 25 Yak Yoghourt samples collected from Gannan pasturing area of Gansu Province. An efficient β-Galactosidase producing strain SYA2 screened out from 21 strains was identified as *Enterobacter sp.* by the means of morphological feature observation, physiological and biochemical characteristics measurement, and 16S rDNA sequence analysis. The β-Galactosidase produced by *Enterobacter sp.* SYA2 was purified by ammonium sulfate precipitation. The optimum temperature, thermostability, optimum pH, and pH stability of the β-Galactosidase, as well as the effects of metal ions on enzymatic activity, were investigated. The optimum temperature of the β-Galactosidase was 40°C. The β-Galactosidase was stable at 35°C and 40°C, but the loss of activity was obvious at 45°C and 50°C. The optimum pH of the β-Galactosidase was 6.5. The β-Galactosidase was stable at pH 6.0~9.0. The β-Galactosidase activity could be activated by Mg^{2+}, Mn^{2+} and Na^+, while inhibited differently by Zn^{2+}, Cu^{2+} and Fe^{2+}. These results also indicate that SYA2 has a more notable effect on lactose hydrolysis in milk.

In: Beta-Galactosidase
Editor: Eloy Kras

ISBN: 978-1-53615-605-8
© 2019 Nova Science Publishers, Inc.

Chapter 1

MICROBIAL BETA-GALACTOSIDASE: PRODUCTION, BIOPROCESS PARAMETERS AND DOWNSTREAM

Tatiane A. Gomes, Michele R. Spier[*]*, PhD*
and Michelle R. F. Vaz, PhD
Chemical Engineering Department, Federal University of Paraná,
Curitiba, Paraná, Brazil

ABSTRACT

β-galactosidase (β-Gal; β-D-galactoside galactohydrolase; EC 3.2.1.23, popularly known as lactase) is an enzyme responsible for catalyzing the hydrolysis of the lactose β-1,4 linkage into α-D-glucose and β-D-galactose. β-galactosidase is found in plants, animals and microorganisms. The immobilized form is widely used in the food industry to reduce the lactose content of dairy products, prevent problems associated with lactose crystallization, and increase sweetness, flavor and

---

[*] Corresponding Author's E-mail: spier@ufpr.br.

solubility. Only microbial β-galactosidase is used for industrial purposes due to its viability and lower cost when compared to other sources. Yeast, filamentous fungi and bacteria are potential producers of intra or extracellular β-galactosidase. Intracellular β-galactosidase is obtained by the permeabilization or disruption of the cell wall membrane after the application of mechanical or non-mechanical methods such as sonication and solvents, respectively. The enzymatic production by microorganisms occurs through submerged fermentation (SmF), in which complex culture media, which are rich in lactose (carbon source) and amino acids (provided by nitrogen sources), are used in the production process; it is necessary to control parameters such as pH and temperature. Both the temperature and optimum pH used in enzymatic production vary according to the microorganism. The same occurs for processes which may, or may not, use agitation and aeration because each microorganism has different conditions of production and oxygen tolerance. In addition to lactose, other components are essential for the catalysis of β-galactosidase, such as metal ions (potassium, manganese, and magnesium), which act as cofactors in enzymatic activity, while glucose, galactose and calcium are known as competitive inhibitors of β-galactosidase. Higher levels of β-galactosidase production are achieved when parameters such as the maximum specific microbial growth rate and yield factors are evaluated and controlled. In addition to the production process, the separation and recovery of enzymes must be controlled due to the high proportion of water in the culture medium and the presence of organic and inorganic molecules, as well as extracellular and intracellular metabolites. The purification process is the most expensive stage of enzyme production, accounting for approximately 80% of the total cost. Downstream can be divided into four general steps: clarification (separation of cells and fragments from the culture medium); low-resolution and high-resolution purifications; and the finishing and packaging of the product. The choice of the purification steps will depend on the application of the biomolecule. For industrial applications, immobilized enzymes are preferable due to their high levels of stability, easy separation from the reaction mixture, and reusability. This chapter discusses the main microorganisms that produce β-galactosidase, the characteristics of the culture media, bioprocessing parameters, and the most relevant downstream steps used in the recovery of microbial β-galactosidase, as well as the main immobilization techniques that are used.

Keywords: biotechnology, downstream, enzyme, immobilization microorganism

INTRODUCTION

Milk and dairy products are consumed by seven billion people worldwide and are considered to be vital sources of nutrition. The average global, per capita milk consumption is approximately 100 Kg milk/year (Food and Agricultural Organization 2010, 2013).

However, lactose intolerance affects many people who are unable to consume these products. Lactose intolerance is caused by the total absence or insufficient production (in childhood) of lactase-phlorizin hydrolase, known as β-galactosidase, or by hypolactasia (in adults). Lactose intolerance can lead to symptoms such as abdominal distension, diarrhea, flatulence and cramps (Heyman 2006; Xavier, Ramana, and Sharma 2018).

β-galactosidase is one of the most important hydrolytic enzymes found in plants, animals, and microorganisms. Its worldwide production is estimated at around 5.75 million metric tonnes/year. This enzyme is a tetramer of four identical polypeptide chains (each monomer contains 1,023 amino acids); it has high specificity for D-galactose and contains lactose as its natural substrate (Fowler and Zabin 1978; Jacobson et al. 1994; Huber and Gaunt 1983; Juers, Matthews, and Huber 2012).

Microbial β-galactosidase is widely used in the food processing industry to produce lactose-free products. It is also incorporated into drugs for lactose-intolerant individuals; used in the production of galactooligosaccharides, to avoid crystallization in food products; and for the treatment of whey and whey permeate, in which it is used to convert the latter into ethanol and sweet syrup (Zhou and Chen 2001; Husain 2010; Saqib et al. 2017; Xavier, Ramana, and Sharma 2018).

Immobilized enzymes are used for industrial applications due to the fact that they have many significant advantages compared to soluble enzymes, such as higher levels of stability and reusability, as well as easier separation of the reaction mixture. which decreases the costs of the process (Grosová, Rosenberg, and Rebroš 2008). Different methods, using a variety of matrices, can be employed to immobilize β-galactosidade. The most common methods include physical adsorption and covalent-binding;

the selection of the most suitable immobilization method depends on the particular enzyme (Tanaka and Kawamoto 1999).

The most common sources of β-galactosidase used in dairy processes are yeast (mainly *Kluyveromyces* sp.) and filamentous fungi (mainly *Aspergillus* sp., which are acceptable as GRAS (generally recognized as safe) by the FDA (Food and Drug Administration). The choice of the enzyme source and type will depend on its application. In addition to the fact that fungal β-galactosidase is extracellular it is more applicable for the treatment of whey due to pH and temperature ranges (acid pH and a temperature range of 25-60°C). β-galactosidase derived from yeast is recommended for lactose hydrolysis processes because it acts in neutral pH; however, because it is intracellular the costs associated with downstream are high (Husain 2010; Saqib et al. 2017; Xavier, Ramana, and Sharma 2018).

Submerged fermentation (SmF) is widely used to produce microbial β-galactosidase; the culture medium contains lactose as carbon sources and other components (nitrogen sources, minerals, and vitamins) with highly controlled pH, temperature, agitation and aeration. At the end of the fermentation process the biomolecules are recovered and concentrated for application in the product (Haider and Husain 2009).

The downstream process is the most expensive stage in the production of microbial enzymes, especially for intracellular enzymes, where permeabilization or cell-disruption methods for enzyme release are required. Moreover, soluble enzymes are unviable for use in the food processing industry due to the high costs involved. Therefore, immobilized enzyme preparations in stirred-tank reactors (STR) are used to improve lactose hydrolysis processes (Haider and Husain 2009; Husain 2010; Xavier, Ramana, and Sharma 2018).

This chapter discusses the different microbial sources for the production of β-galactosidase and aspects of production such as carbon and nitrogen sources, as well as the main bioprocessing parameters. The main methods for cell disruption and most relevant steps in the downstream process and immobilization are also presented.

LACTOSE

Lactose (4-O-β-D-galactopyranosyl-D-glucopyranose) is a reducing disaccharide formed by a glucose molecule linked to a galactose molecule. It comprises about 52% of the total non-fat solids present in milk and about 70% of the solids present in whey; it has a melting point of 201.6°C. In dairy products it is present in two non-crystalline forms (α-hydrated and β-anhydrous) and it can also be present in mixtures thereof in equal proportions (Gänzle, Haase, and Jelen, 2008; Haider and Husain 2009; Damodaran, Parkin, and Fennema 2008; Harju, Kallioinen, and Tossavainen 2012; Mattar, Mazo, and Carrilho 2012; Klein et al. 2013; Nivetha and Mohanasrinivasan 2017).

In supersaturated solution at temperatures lower than 93.5°C lactose crystallizes as anhydrous β-lactose, which is sweeter than the monohydrate form (α-hydrated). At temperatures above 93.5°C lactose crystallizes in the form of α-lactose monohydrate: α-lactose may be transformed into β-lactose by the process of mutarotation at the equilibrium (37.3% α-lactose and 62.7% β-lactose). The growth of lactose crystals may be influenced by factors such as the degree of supersaturation and the presence or absence of inhibitors (Damodaran, Parkin, and Fennema 2008).

LACTOSE INTOLERANCE

Lactose intolerance, or the maldigestion of lactose, is a reduction in the ability to hydrolyze this disaccharide (hypolactasia). Approximately 75% of the world's population could lose this ability, while others can digest lactose into adulthood (Mattar, Mazo, and Carrilho 2012).

Hypolactasia can be classified into three types: congenital (a rare phenomenon that is genetically determined as the total inability of the organism to produce the enzyme); primary (the most common type, characterized by low levels of enzymatic activity); and secondary, which is temporarily caused by infectious enteritis (i.e., giardiasis) or mucosal damage due to coeliac disease, inflammatory bowel disease, drugs, gastrointestinal surgery or short bowel syndrome, all of which lead to either a reduction in absorptive capacity or the downregulation of lactase expression in the small intestine (Heyman 2006; Mattar, Mazo, and Carrilho 2012).

β-galactosidase activity in humans originates from the lactase-phlorizin hydrolase enzyme (popularly known as lactase), which is produced in the brush border of the small intestine; its activity is confirmed on the surface of epithelial cells (Strzalkowska, Jasinska, and Jozwik 2018).

Lactose absorption occurs by the hydrolysis of lactose in glucose and galactose within intestinal enterocytes. These monosaccharides are easily assimilated in the bloodstream. Glucose absorption is initially used as the main, vital source of energy, while galactose is used as a glycolipid and glycoprotein component (Nivetha and Mohanasrinivasan 2017).

Unabsorbed lactose remains in the large intestine, which leads to an increase in osmotic load in the lumen fluid and electrolytes. Intestinal bacteria use lactose as a substrate for their growth, leading to symptoms such as flatulence due to the production of volatile fatty acids and gases by bacterial metabolism, in addition to distension, abdominal cramps, and diarrhea (de Vrese et al. 2001; Heyman 2006; Jellema et al. 2010; Levitt, Wilt, and Shaukat 2013; Misselwitz et al. 2013; Nivetha and Mohanasrinivasan 2017).

BETA-GALACTOSIDASE

β-galactosidase (lactase, galactohydrolase, β-D-galactoside-galactohydrolase, EC 3.2.1.23) is responsible for catalyzing the hydrolysis of the β-galactopyranosyl terminal residue of lactose (Galβ1-4Glc) to form glucose and galactose (Clark and Blanch 1997; Holsinger 1997). This enzyme can be found in animals (mammalian tissues, e.g., fish) plants and microorganisms. In mammals, β-galactosidase activity is used to quantify replicative senescence (loss of cell division capacity) in vitro and cell aging in vivo (Demirhan, Apar, and Özbek 2010).

During the last ten years studies have been carried out based on β-galactosidase activity in order to monitor the cellular aging of fish and cell characterization (Vo et al. 2015; Sakai et al. 2017).

In humans, enzymatic activity has been studied to verify cellular senescence in intervertebral tissue by measuring the percentage of cells with senescence (Gruber et al. 2007).

β-galactosidase is naturally found in plants it and it contributes to their growth and ripening (Saqib et al. 2017). A study of persimmon (Diospyros kaki) verified that fruit ripening occurred due to the loss of galactose residues from the cell wall by β-galactosidase that was synthesized by the plant itself. β-galactosidase catalyzes the cleavage of terminal residues of D-galactose at β-1,4 linkage, as occurs in other fruits such as apples, pears, avocado, strawberry, papaya and kiwi (Kang et al. 1994; Husain 2010; Saqib et al. 2017).

According to Haider and Husain (2007), β-galactosidase from almond (*Amygdalus communis*) that was precipitated by ammonium sulfate presented optimum pH and temperatures of 5.5 and 50°C, respectively. It was shown to be significantly stable in relation to heat, pH, calcium and magnesium ions, and D-galactose.

Even with high levels of stability, there are some disadvantages associated with the use of plant tissues as sources of enzymes, such as the fact that they present supply limitations, need large area crops, are subject to bad weather, and the plant tissue from which the enzyme has been extracted cannot be reused. Therefore, at the industrial level, only enzymes of microbial origin are used because they are more viable and of lower cost than those derived from animal and plant sources.

MICROBIAL BETA-GALACTOSIDASE

Bacteria, yeasts and filamentous fungi are all producers of β-galactosidase. Depending on the microorganism, microbial

Table 1. Main producers of β-galactosidase and conditions for optimal activity (pH and temperature)

	Microorganism	pH	Temperature (°C)	Reference
Filamentous fungi	Aspergillus oryzae	4.5 - 4.8	35 - 50	Husain (2010); Freitas et al. (2011); Vera, Guerrero, and Illanes (2011); Ansari and Husain (2012); Fischer et al. (2013); Benjamins et al. 2014)
	Penicillium sp.	-	37	Silvério et al. (2018)
	A. niger	3.0 - 4.0	55 - 60	Gekas and Lopez-Leiva (1985)
	A. flavus	-	-	Gekas and Lopez-Leiva (1985)
	A. foetidus	-	-	Shukla and Wierzbicki (1975)
	A. phoenicis	4.0 - 5.5	40 - 55	Gekas and Lopez-Leiva (1985); Shukla and Wierzbicki (1975)
	Mucor pusillus	4.5 - 6.0	60	Gekas and Lopez-Leiva (1985)
	M. meihei	-	-	Gekas and Lopez-Leiva (1985)
	Neuspora crassa	4.5 - 7.0	40 - 50	Gekas and Lopez-Leiva (1985)
	Terathosphaeria acidotherma	2.0 - 4.0	25	Isobe et al. (2013)
	Fusarium moniliforme	3.8 - 5.0	50 - 60	Macris and Markakis (1981)
	Trichoderma reesei	5.0	28	Gekas and Lopez-Leiva (1985)
	Alternaria palmi	4.0	25 - 30	Agrawal, Sonawat, and Dutta (1982)
	Curvularia inaequalis	3.7 - 4.5	60	Zagustina and Tikhomirova (1976)
	Scopulariopsis sp.	5.0	45	Santos, Simiqueli, and Pastore (2009)
Yeast	Kluyveromyces fragilis	6.6	37	Gekas and Lopez-Leiva (1985)
	K. lactis	6.9 - 7.3	35	Dagbagli and Goksungur (2008); Shen et al. (2012); Gomes et al. (2018)
	K. marxianus	5.0 - 5.2	30	Gekas and Lopez-Leiva (1985)
	Candida pseudotropicalis	6.0 - 7.5	30 - 40	Gekas and Lopez-Leiva (1985)
	C. kefyr	6.2	45 - 47	Gekas and Lopez-Leiva (1985)
	Torulopsis versatilis	6.0 - 7.5	30 - 40	Gekas and Lopez-Leiva (1985)
	T. sphaerica	6.0 - 7.5	30 - 40	Gekas and Lopez-Leiva (1985)

	Microorganism	pH	Temperature (°C)	Reference
Bacteria	Escherichia coli	7.2	40	Gekas and Lopez-Leiva (1985)
	Streptococcus thermophilus	6.5 - 7.5	55	Gekas and Lopez-Leiva (1985)
	S. fragilis	-	-	Gekas and Lopez-Leiva (1985)
	S. cremoris	6.0 - 7.0	30 - 50	Gekas and Lopez-Leiva (1985)
	S. mitis	6.0 - 65	30 - 40	Campuzano et al. (2009)
	Bacillus sp.	6.0	60	Gekas and Lopez-Leiva (1985)
	Lactococcus lactis	7.0	37	Li et al. (2012)
	Lactobacillus helveticus	6.0 - 7.0	30 - 50	Gekas and Lopez-Leiva (1985)
	L. reuteri	6.5	37°C	Hidalgo-Morales, Robles-Olvera, and Garcia (2005); Gomes et al. (2018)
	L. bulgaricus	6.5 - 7.0	42 - 45	Ustok, Tari, and Harsa (2010); Gomes et al. (2018)
	L. thermophilus	6.2 - 7.1	55 - 57	Gekas and Lopez-Leiva (1985)
	L. plantarum	6.5 - 7.5	50	Selvarajan and Mohanasrinivasan (2015); Gomes et al. (2018)
	L. acidophilus	6.5 - 7.5	45	Andersen et al. (2011); Carevic et al. (2015); Gomes et al. (2018)
	L. fermentum	6.5 - 7.0	40	Liu et al. (2011); Gomes et al. (2018)
	L. johnsonii	6.5	37	Gomes et al. (2018)
	Pediococcus pentosaceus	6.5	37	Gomes et al. (2018)
	P. acidilactici	6.5	37	Gomes et al. (2018)

β-galactosidase has diverse functional properties. β-galactosidase derived from fungi exhibit optimal activity in acid pH and temperatures ranging from 25 to 60°C. Those produced by yeasts have their optimal activity at higher pH levels, and temperatures between 30 and 40°C, depending on the genus and species of the yeast. β-galactosidases derived from bacteria show optimal activity at pH levels around neutrality (between 6.0 and 7.5) and temperatures between 30 and 50°C. The main producers of β-galactosidase, as well as the optimal pH levels and temperatures, are shown in Table 1.

Thus, microorganisms are potential sources of enzymes because their production can be increased quickly. Although microbial enzymes are highly sensitive to genetic alterations, the use of controlled conditions and molecular biology techniques of make it possible to create improved strains in terms of the production and quality of the required enzymes, which are characterized by high productivity and lower cost when compared to the production of enzymes from vegetable sources (Couri and Damaso 2015).

As previously mentioned, microbial enzymes do not need large crop areas for production. Generally, this process occurs in fermenter flasks or bioreactors with control of the process parameters, and the time necessary for biosynthesis is only a few days.

β-galactosidase derived from *Kluyveromyces* sp. and *Aspergillus* sp. is the enzyme that is most used in dairy processes on an industrial scale. *Kluyveromyces lactis* is an important source of β-galactosidase because it is found naturally in dairy products. Furthermore, its pH range (6.9 to 7.3, see Table 1) means that it can be used in the production of lactose-free milk (Rech and Ayub 2007; Husain 2010). In yeasts, this enzyme is produced intracellularly, and disruption methods (See section on cell disruption later in this chapter) are required for its release. β-galactosidase derived from *Aspergillus oryzae* is also widely used in the dairy industry; it is an extracellular and thermostable enzyme (37 to 50°C, see Table 1). However, due to its pH range (4.5 to 4.8, see Table 1) it is most effective for lactose hydrolysis in acidic products such as whey (Husain, 2010).

Lactic acid bacteria have been studied in relation to β-galactosidase production due to their health and technological benefits, as well as the fact that they are potential producers of β-galactosidase. The advantages of these microorganisms as sources of enzymes are their GRAS status, as well as the fact that the probiotic activity of some strains, which improves lactose digestion, means that they can be applied without extensive purification steps (Vasiljevic and Jelen 2002; Vinderola and Reinheimer 2003; Ansari and Husain 2010; Gomes et al. 2018).

Another reason for the utilization of lactic acid bacteria as a source of β-galactosidase is the production of thermostable enzymes. *Lactobacillus delbrueckii* ssp. *bulgaricus* and *Streptococcus thermophilus* are distinguished by the production of an enzyme with high stability at temperatures of approximately 50-60°C and high levels of enzymatic activity (Tari, Ustok, and Harsa 2009; Choonia and Lele 2011; Abbasalizadeh, Hejazi and Hajiabbas 2015).

The synthesis of β-galactosidase was demonstrated using *Escherichia coli* as a model microorganism. All the genes involved in this synthesis are located close to each other and together they form the *lac operon*, which consists of three structural genes: *lacZ*, *lacY*, and *lacA*. The *lacZ* gene codes for β-galactosidase (the enzyme responsible for the hydrolysis of lactose); *lacY* codes for permease (the enzyme responsible for the transport of lactose through the microbial cell wall) and, *lacA* codes for transacetylase (Dragosits et al. 2014; Nath et al. 2014).

The functional properties of microbial β-galactosidase can be improved by recombinant DNA techniques, which are used to express and optimize enzyme production from different sources. This technique can be used for microbial enzyme production, including different microbial sources which are not GRAS (Xavier, Ramana, and Sharma 2018).

The high level of interest in the study and application of recombinant enzymes is due to the factors such as their ease of use in purification and large-scale production, their rigidity and permeability, their regenerability (which is due to the stability caused by the protection of their active sites from deactivation), their hydrophobic/hydrophilic character, and the fact that they are easy to separate from the reaction mixture without the use of chemicals or heating (Ansari and Satar 2012).

LACTOSE HYDROLYSIS

The enzymatic hydrolysis of lactose is one of the most important mechanisms used in the food industry. According to Mahoney (1985) and Zhou and Chen (2001), the active site of β-galactosidase contains cysteine, which acts as a donor of protons, and histidine, which serves as an acceptor of protons. The sulfhydryl group of cysteine acts as a proton donor, while the imidazole group of histidine acts as a nucleophilic site, facilitating the cleavage of the glycosidic bond during the enzymatic hydrolysis of lactose. This mechanism was described using lactase derived from *E. coli* by Wallenfels and Malhotra (1960).

However, many authors have reported that β-galactosidase derived from different microbial sources, such as bacteria, yeast and filamentous fungi, contains two glutamic acid residues (GLU482 and GLU551), which simultaneously act as a donor of protons and a nucleophilic site, respectively, during the enzymatic reaction. Therefore, the action mechanism of β-galactosidase in the hydrolysis of lactose consists of the formation of the enzyme-galactosyl complex and the simultaneous release of glucose (López Leiva and Guzman 1995; Raymond R. Mahoney 1998; Rustom, Foda, and López-Leiva 1998; Sheu et al. 1998)

APPLICATIONS OF BETA-GALACTOSIDASE

β-galactosidase has been studied due to its nutritional and technological benefits. In the food industry it is used in the production of milk with reduced lactose content or lactose-free milk, lactose-free dairy products (such as yogurts, sorbets, condensed milk, sour cream and cheeses). It is also incorporated into drugs for lactose-intolerant individuals and used in the production of food syrups, as well as being incorporated in the treatment of whey and whey permeate, and the production of galactooligosaccharides. Furthermore, oral administration of the microbial enzyme is recognized as minimizing the side effects of lactose intolerance.

The medicines for lactose-intolerant individuals that are commercially available contain β-galactosidase derived from fungi, normally *Aspergillus* sp., which is stable at low pH levels, allowing for proper functioning in the stomach (Panesar et al. 2007; Saqib et al. 2017).

Studies regarding the oral administration of microbial β-galactosidase in lactose-intolerant individuals has proved that the side effects of intolerance were minimized (Li et al. 2012; Dominguez-Jimenez et al. 2014; de Vrese et al. 2015; Roškar et al. 2017). The aforementioned authors verified that the oral administration of microbial lactase produced by a single microorganism, or enzymes from different microbial sources, increased lactose digestion and minimized symptoms such as diarrhea, abdominal pain and flatulence in individuals with an absence of intestinal lactase activity.

The treatment of milk with enzymes occurs in bioreactors with controlled pH levels and temperatures. Due to the cost of these processes, techniques can be used such as soluble enzyme application, with subsequent recycling by membrane processes or immobilized enzymes (Husain 2010).

For industrial applications, immobilized β-galactosidase (see the 'Immobilization of microbial beta-galactosidase' section later in this chapter) is preferable because it retains biochemical activity, has higher levels of stability during operation and storage, and is easily removed from the reaction mixture when compared to free enzymes, therefore enabling it to be reused (Husain 2010).

β-galactosidase can also be used to reduce crystallization in ice cream and condensed milk, which occurs due to high lactose concentrations. Therefore, the enzyme is used to hydrolyze the lactose in these products and improve their texture (Saqib et al. 2017).

Another application is in the production of galactooligosaccharides (GOS) using the process of transglycosylation, which occurs during lactose hydrolysis. GOS are important prebiotics that are used as ingredients in foods due to their health benefits such as resistance to gastric acidity, hydrolysis and intestinal absorption, fermentation by intestinal microflora, and the selective stimulation of the growth and/or activity of intestinal bacteria associated with health and wellbeing (Gibson et al. 2004; Maischberger et al. 2010).

In addition to these applications, β-galactosidase can also be used in the treatment of whey permeate. After whey protein is extracted from whey by ultrafiltration the whey permeates (which are rich in lactose, salts, minerals, nitrogen, non-proteins and water) are discarded as dairy effluent because they can cause pollution when dropped into water streams due to their high biochemical and chemical oxygen demands (BOD and COD). The use of β-galactosidase in subsequent treatment can convert whey and whey-based products into substrates for cell cultivation (Parashar et al. 2016; Saqib et al. 2017). Lactose hydrolyzed from whey may also be used to produce sweetening syrups utilized in ice cream, and bakery and confectionary products (Saqib et al. 2017; Xavier, Ramana, and Sharma 2018).

COMPOSITION OF CULTURE MEDIUM AND BIOPROCESS PARAMETERS IN MICROBIAL BETA-GALACTOSIDASE PRODUCTION

Microbial enzymes can be produced by different fermentation processes such as solid, semi-solid and submerged fermentations.

Solid-state fermentation (SSF) is defined as a process in which there is the absence, or almost absence, of free water in the culture medium. However, the substrate must have enough moisture to guarantee the development of microorganisms (Pandey 1992; Singhania et al. 2009). In addition to the possibility of using agroindustrial waste as a substrate, SSF processes require low energy and generate less wastewater (Pandey 2003).

In some cases, excess substrate may cause the inhibition of microbial growth; this has been observed in relation to SSF using plants or residues with a high concentration of sugars. When this occurs the operation system can be modified by increasing the water content to improve the availability of substrate. This system is called semisolid-state fermentation (Economou et al. 2010).

Another fermentation process that is widely used to produce biomolecules is submerged fermentation (SmF), or submerged culture (SmC), which occurs in a medium with free water. This process in most widely used to produce bioproducts such as antibiotics, alcohol, organic acids, vitamins and enzymes (Subramaniyam and Vimala 2012; Hansen et al. 2015). It is also widely used to produce microbial β-galactosidase.

In this process there are no problems regarding the diffusion of components due to the homogeneity of the medium (Raj and Karanth 2006). SmF has several advantages when compared to other processes such as heat and mass transfers, greater efficiency, and easier control of fermentation. In this process, the high concentration of water avoids degrading the product, since it facilitates the control of temperature (Raj and Karanth 2006; Rao 2009).

Enzymatic hydrolysis of lactose, which is used to produce milk, dairy products or pure lactose, is performed using bioreactors, usually by the batch process with highly controlled pH and temperature conditions. These conditions vary depending on the microbial source of β-galactosidase (Husain, 2010). Stirred-tank reactors (STRs) are commonly used for commercial applications of lactose hydrolysis due to the ease of process control (Haider and Husain, 2009).

CARBON AND NITROGEN SOURCES

Lactose in the culture medium induces the production of β-galactosidase; microorganisms capable of producing this enzyme hydrolyze lactose into glucose and galactose (Akcan 2011; Alikkunju et al. 2016). Culture medium with lactose as a carbon source, provided by milk,

milk whey and cheese whey, is used for microbial enzymatic production (Arukha et al. 2015).

Arukha et al. (2015) studied different carbon sources for β-galactosidase production using *Lactobacillus fermentum* in MRS (Man Rogosa and Sharpe broth) broth, with pH of 6.5 and supplemented with glucose, fructose, galactose, sucrose and maltose. An additional study was performed of glucose (30 and 60 mM) in MRS supplemented with galactose (30 mM), using 2% (v/v) inoculum. The authors verified that the enzymatic activity was approximately 2.0 U mL^{-1} in the MRS medium with glucose. Similar activity was found in the MRS medium with lactose, galactose and sucrose. A reduction of 8.5% was found in the MRS with maltose (1.83 U mL^{-1}), and the culture medium with fructose showed the lowest enzymatic activity (1.24 U mL^{-1}), i. e., enzymatic activity that was 38% lower than the MRS with glucose. β-galactosidase activity was not repressed when glucose (30 and 60 mM) was used, and when MRS supplemented with galactose (30 mM) was combined.

However, studies by Hsu, Yu, and Chou (2005) and Carevic et al. (2015) demonstrated that lactose improved β-galactosidase activity, and in a culture medium with glucose enzymatic activity was repressed. Therefore, glucose may act as a competitive inhibitor of the enzyme (Nguyen et al. 2006).

The same was verified by Ansari and Husain (2012) when studying the effect of increased galactose/glucose concentrations (1.0 – 5.0% m/v) on β-galactosidase activity by *A. oryzae* in 0.1 M sodium acetate buffer (pH 4.5 and 37°C). The authors used a control sample without added monosaccharide and found enzyme activity of 2.0 U mL^{-1}. It was observed that concentrations of 5.0% (m/v) galactose and less than 2.0% (m/v) glucose inhibited β-galactosidase activity, resulting in a 70% decrease in initial activity. Therefore, based on previous studies, it is possible to verify that lactose promotes an increase in enzymatic activity, while glucose and galactose are potential inhibitors.

As with carbon sources, the use of nitrogen sources is essential for the production of β-galactosidase. Organic and inorganic nitrogen sources can be used, such as yeast extract, meat extract and peptone, as well as from various agroindustrial residues (cheese whey and corn steep water). The most widely used inorganic nitrogen sources are ammonium nitrate, ammonium chloride, ammonium sulfate and sodium nitrate (Akcan 2011; Fischer and Kleinschmidt 2015; Raol et al. 2015; Gomes et al. 2018).

Manera et al. (2011) studied the optimization of β-galactosidase production by *Kluyveromyces marxianus* CCT 7082, using agroindustrial residues (corn steep water, cheese whey and commercially available yeast hydrolysate) as nitrogen sources and salts (magnesium sulfate and potassium phosphate monobasic). Different concentrations of lactose were tested using different nitrogen sources: 10 - 70 g L^{-1} in cheese whey; 10 - 100 g L^{-1} in corn steep water; and 4 - 20 g L^{-1} in yeast hydrolysate. The pH was between 5.0 and 7.0. The parameters that resulted in the highest enzymatic activity and productivity (1,400 U g^{-1} and 61 U L h^{-1}, respectively) were 70 g L^{-1} lactose, 65 g L^{-1} of corn steep water and 4 g L^{-1} of yeast hydrolysate at pH 5.0.

Other nitrogen sources, such as casein, albumin, partially demineralized whey, hydrolyzed collagen, soybean extract and inactive beer yeast, were also studied in relation to β-galactosidase production by *Lactobacillus reuteri* B-14171 (Gomes et al. 2018). This study showed that casein and inactive beer yeast combined (3.0 g L^{-1} of each) in modified MRS medium (MMRS with 20 g L^{-1} lactose instead of glucose) resulted in the highest enzymatic activity (1,269 U L^{-1}). The lowest enzymatic activity was found in the MMRS with soybean extract (4.0 g L^{-1}). The authors attributed the variation in enzymatic activity to the amino acid composition of the nitrogen sources that were tested. Casein and inactive beer yeast have high concentrations of glutamic and aspartic acids and cysteine, which were reported by Akcan (2011) as being essential for enzyme production. These amino acids are also found in commercial yeast extract, making it the most commonly used nitrogen source in enzymatic production. However, alternative nitrogen sources have been studied to reduce production costs (Bansal et al. 2008; Machado et al. 2015; Cardoso et al. 2017; Gomes et al. 2018).

Thus, several nitrogen sources can be used in culture medium to produce β-galactosidase; their choice depends not only on the amino acid composition but also on the microorganism that is used.

COFACTORS FOR BETA-GALACTOSIDASE PRODUCTION

In addition to the substrate, other components, which are often referred to as cofactors, may be essential for enzymatic catalysis. These substances can be classified into two groups: specific coenzymes (compounds of low molecular mass and complex structure) which act in the transport of specific chemical groups and; activators (usually metallic ions) that serve in the formation of the activated complex without participating in the reaction (Nelson, Cox, and Lehninger 2013).

There are also many substances that act as inhibitors of enzymatic activity, such as urea, which are considered non-specific because they act by inhibiting a wide variety of enzymes; others only act in relation to a specific group of enzymes and are referred to as inhibitors of enzymatic activity (Chaplin and Bucke 1990).

Enzymatic inhibition may be reversible or irreversible. Reversible inhibition commonly occurs through a competitive inhibitor which competes with the substrate for the same active site of the enzyme and is structurally related to the substrate. Non-competitive inhibitors bind to the enzyme or to the enzyme-substrate complex, i.e., they do not bind to the active site of the enzyme but to another catalytically inactive site, thereby altering the Michaelis-Menten constant value. In the process of irreversible inhibition, an inhibitor binds covalently or destroys a functional activity that is essential for enzyme activity (Chaplin and Bucke 1990; Nelson, Cox, and Lehninger 2013).

According to Ansari and Husain (2010), galactose is a potent inhibitor in relation to β-galactosidase because it competes for the active site of the enzyme. However, Guven et al. (2011) studied the effects of cofactors and inhibitors of β-galactosidase by *Alicyclobacillus acidocaldarius* ssp. *rittmannii*, and found that galactose can act as a mixed-type inhibitor (competitive and non-competitive), resulting in changes in K_m and V_{max} values, as a decrease in enzyme-substrate affinity and in the maximum reaction rate (Segel 1979). In this type of inhibition, the inhibitor also binds to a site other than the active site where the substrate binds (Guven et al. 2011).

Glucose can act as a non-competitive inhibitor in species such *as L. reuteri* and *Bifidobacterium longum* BCRC 15708 (Nguyen et al. 2012). However, Mateo et al. (2004) reported that glucose can act as a competitive inhibitor in species such as *K. lactis* and *Thermus* sp., causing a decrease in the rate of reaction or even stopping it altogether.

Metal ions are required to activate some β-galactosidases; the latter can be added as salts and their effect depends on the concentration that is applied.

In general, fungal β-galactosidases are known to be less dependent on the effects of metal ions (Mlichova and Rosenberg 2006; Liu et al. 2015; Cardoso et al. 2017). However, according to Juers, Matthews, and Huber (2012), some β-galactosidases require sodium or potassium and magnesium to be fully active. These ions are essential for binding and reactivity; however, some residual activity can be found in their absence.

Nguyen et al. (2006) evaluated the effects of mono and divalent ions on β-galactosidase activity from *L. reuteri*. It was verified that the enzyme was activated by Na^+, K^+, Mn^{2+} and Mg^{2+}; in contrast, Ca^{2+} caused the inhibition of β-galactosidase activity. Different behavior was found by Cardoso et al. (2017) regarding β-galactosidase activity from *Aspergillus lacticoffeatus*. The authors verified that neither mono nor divalent ions promoted a significant enhancement of hydrolytic activity. Moreover, Li^+, K^+, Ba^{2+} and Fe^{2+} promoted an adverse effect on enzymatic activity. Therefore, enzymatic activity is affected in different ways depending on the enzyme source.

OTHER PARAMETERS
(pH, TEMPERATURE, AGITATION AND AERATION)

Temperature and pH are physicochemical parameters that influence the production of microbial β-galactosidase; both parameters (optimum temperature and pH) vary according to the microorganism producer. The same occurs for processes that use (or don't use) agitation and aeration since each microorganism has different conditions of production and oxygen requirements. Gentle agitation of the culture medium favors the mass transfer and dissolution and homogenization of the culture medium components, providing the cells access to the nutrients necessary for their development and the synthesis of biomolecules, which are essential for aerobic microorganisms.

Alves et al. (2010) observed the effect of agitation and aeration on the increase in β-galactosidase production by *K. marxianus* CCT 7082. A stirred fermenter, with agitation varying between 200 and 500 rpm (rotations per minute) and aeration rate between 0.5 and 1.5 vvm (volume of air per volume of culture medium per minute), were used. The authors observed that the production and enzymatic activity were strongly influenced by aeration and agitation: the optimal conditions were 500 rpm and 1.5 vvm during 14 hours fermentation, with enzymatic activity and productivity of 17.0 U mL^{-1} and 1.2 U mL^{-1}, respectively.

Table 2 shows the optimum pH and temperature, type of process used (with agitation or static, and with or without aeration), as well as the fermentation time, in the production of β-galactosidase from different microbial sources reported in the literature.

Process conditions depend on the species of the microorganism that is studied since the requirements vary not only according to the species, but also between the same genus and the substrate used for enzymatic production. β-galactosidase derived from filamentous fungi and yeast requires agitation speed for its production, while enzymes from bacteria

Table 2. Bioprocess parameters of β-galactosidase production from different microbial sources

	Microorganism	Culture medium	pH	Temperature (°C)	Fermentation process conditions	Time (h)	Reference
Filamentous Fungi	Aspergillus oryzae	Cheese whey solution (5% (w/v)	5.0	28	Agitation (120 rpm)	120	Viana et al. (2018)
	A. acticoffeatus	Synthetic medium*	6.5	28	Agitation (120 rpm)	144	Cardoso et al. (2017)
	A. brasiliensis	Synthetic medium*	6.5	28	Agitation (150 rpm)	480	Silvério et al. (2018)
	A. uvarum	Synthetic medium*	6.5	28	Agitation (150 rpm)	480	Silvério et al. (2018)
	A. restrictus	Synthetic medium*	6.5	28	Agitation (150 rpm)	480	Silvério et al. (2018)
	Penicillium brevicompactum	Synthetic medium*	6.5	28	Agitation (150 rpm)	480	Silvério et al. (2018)
	P. italicum	Synthetic medium*	6.5	28	Agitation (150 rpm)	480	Silvério et al. (2018)
	P. spinulosum	Synthetic medium*	6.5	28	Agitation (150 rpm)	480	Silvério et al. (2018)
	Mucor sp.	Synthetic medium*	6.5	28	Agitation (150 rpm)	480	Silvério et al. (2018)
	Trametes versicolor	Synthetic medium*	6.5	28	Agitation (150 rpm)	480	Silvério et al. (2018)
Bacteria	Lactobacillus bulgaricus	Broth of skimmed milk (8 – 16% w/v)	6.78	35 - 43	Stationary	8	Tari, Ustok and Harsa (2009)
	Streptococcus thermophilus	Whey and corn steep liquor (0 – 5% w/v of each), peptone and potassium phosphate (2% w/v of each),	7.2	35 - 45	Stationary	8	Tari, Ustok, and Harsa (2009)
	L. acidophilus	MRS + lactose (10 g L^{-1})	6.5 – 7.5	45	Stationary	24	Choonia and Lele (2011)
	L. plantarum	Lactose medium*	-	25	Stationary	24	Selvarajan and Mohanasrinivasan (2015)
	L. reuteri	MRS + lactose (1% w/v) MMRS + casein and inactive beer yeast (3 g L^{-1} of each)*	6.5	37	Stationary	22; 48	Nguyen et al. (2006); Gomes et al. (2018)
	Bacillus licheniformis	Laura broth liquid medium*	7.0	37	Agitation (150 rpm)	120	Akcan (2011)

Table 2. (Continued)

	Microorganism	Culture medium	pH	Temperature (°C)	Fermentation process conditions	Time (h)	Reference
Yeast	*Kluyveromyces lactis*	SDB – lactose (10 g L^{-1}); fermentation medium*	5.6; 8.45	28; 27.6	Agitation: 250 rpm; 200 rpm with 0.75 vvm aeration and 0.1 MPa (tank pressure)	48; 12	Dagbagli and Goksungur (2008); You et al. (2017)
	K. marxianus	Medium described by Manera et al. (2008); medium (% w/v): lactose (10) + peptone (1) + yeast extract (0.5) + chloramphenicol (0.01)	6.0; 3.0	30; 20	500 rpm and 1.5 vvm; 250 rpm	14; 64	Alves et al. (2010); Al-jazairi et al. (2015)

*Synthetic medium: (% w/v): lactose (2.0), peptone (0.4), yeast extract (0.4), KH_2PO_4 (0.2), $Na_2HPO_4 \cdot 12H_2O$ (0.8) and $MgSO_4 \cdot 7H_2O$ (0.025).
*MMRS: modified Man, Rogosa and Sharpe broth with 20 g L^{-1} of lactose instead of glucose.
*Lactose medium: lactose (10 g L^{-1}), yeast extract (10 g L^{-1}), protease peptone (10 g L^{-1}), magnesium sulfate (50 mg L^{-1}), manganese sulfate (25 mg L^{-1}), tri-ammonium citrate (4 g L^{-1}), potassium acetate (2.5 g L^{-1}) and dipotassium hydrogen phosphate (4 g L^{-1}).
*Laura broth liquid medium (% w/v): yeast extract (1), peptone (0.5) and sodium chloride (0.5).
*Fermentation medium (g L^{-1}): lactose (20), yeast extract (3), malt extract (3) and peptone (3).
*Medium described by Manera et al. (2008) (g L^{-1}): NH_4SO_4 (0.8), yeast malt (17.0), KH_2PO_4 (5.0), $MgSO_4 \cdot 7H_2O$ (0.4) and lactose (28.8).

KINETIC PARAMETERS

The evaluation of kinetic parameters is essential to determine the concentrations of one or more components that are part of the system during fermentation such as microorganism (biomass); metabolites (products) and substrate (nutrients for growth) (Uzir and Don 2008).

According to Katoh, Horiuchi and Yoshida (2015), bioprocessing involves many chemical and/or biochemical reactions and it also requires knowledge about the ratio between reactants (substrates) and products. In addition, substrate consumption rates and bioproduct formation throughout the process become important, not only for the control of bioprocessing but also in the choice of bioprocessing type (batch or fed-batch, for example). Knowledge regarding this information makes it possible to obtain parameters such as the specific velocity of substrate consumption, biomass and product formation, maximum specific growth rate, yields or conversion factors, and process productivity.

MAXIMUM SPECIFIC GROWTH RATE (μ_{MAX})

The growth of most microbial cells consists of a hyperbolic curve, where the cell growth rate is affected by pH, temperature, composition of the culture medium, substrate concentration, oxygen supply rate and other parameters. When these parameters are constants, the maximum specific growth rate could be affected by limiting substrates.

Lactose is the limiting substrate for β-galactosidase production. In a study of the β-galactosidase derived from *Bacillus safensis* using whey, Nath et al. (2016) found that the lactose concentration influenced microbial cell growth and, consequently, the production of β-galactosidase. Based on mathematical models such as Monod and Luedeking-Piret, the authors confirmed that high initial lactose concentrations (> 20 g L^{-1}) in the culture medium might signify inhibition of the production of intracellular β-galactosidase.

Abbasalizadeh, Hejazi, and Hajiabbas (2015) evaluated the effects of different culture media on the specific (μ) and maximum growth rate (μ_{max}) of *L. bulgaricus*. The culture medium constituted whey, whey protein concentrate, casein hydrolysate and minerals; it had a concentration of lactose of 39.1 g L^{-1} lactose and presented the highest μ and μ_{max} values (0.77 h^{-1} and 1.17 g L^{-1}, respectively). Although the lactose concentration was not the highest of the culture media that were tested, the highest μ_{max} was due to the supplementation of the culture medium with readily available nitrogen sources. Therefore, in addition to the substrate, other components of the culture medium influence the microbial growth rate, as well as yield factors and productivity.

Experimental data are used to determine growth parameters. For this purpose, viable cell counts as colony forming units (CFU mL^{-1}) by surface plate dilutions are carried out and incubated at a controlled temperature. The culture media and temperature conditions are variable depending on the microorganism. The total biomass concentration can be carried out by the thermogravimetric technique or by the optical density measurements, which are converted to biomass dry weight using a calibration curve.

YIELD FACTORS

Microbial growth, as well as product formation, are bioconversion factors, where the nutrients that are part of the culture medium during the fermentation process are converted into cell concentrations and metabolites. These conversions can be quantified in yield coefficients that represent the efficiency of the fermentation process. The yield factors relate to the values of X (biomass), S (substrate) and P (product) over time (Chaplin and Bucke 1990; Shuler and Kargi 2002).

Table 3. Yield factors reported in the literature for β-galactosidase production using different culture media and microbial sources

Yield factor*	Value	Microorganism	Culture medium and growing conditions	Reference
$Y_{P/X}$ (U g^{-1})	480	*K. marxianus*	Medium with 2% lactose (pH 5.5) and 150 rpm	Rajoka, Samia, and Sahid (2003)
	590	*K. marxianus*	Medium by Manera et al. (2008)** and 350 rpm + 1 vvm	Alves et al. (2010)
$Y_{P/S}$ (U g^{-1})	66.59	*Lactobacillus delbrueckii* ssp. *bulgaricus* and *Streptococcus thermophilus*	Culture media A and B***	Tari, Ustok, and Harsa (2009)
	82.32	*L. reuteri*	MMRS (2% lactose)****	Gomes et al. (2018)
	16.94	*K. lactis*	MSDB (4% lactose)****	Gomes et al. (2018)
	66.59	*L. delbrueckii* ssp. *bulgaricus* and *S. thermophilus*	Culture media A and B***	Tari, Ustok, and Harsa (2009)
$Y_{X/S}$ (g g^{-1})	0.35	*K. marxianus*	Medium by Manera et al. (2008)** and 500 rpm + 1.5 vvm	Alves et al. (2010)
	0.353	*L. delbrueckii* ssp. *bulgaricus* and *S. thermophilus*	Culture media A and B***	Tari, Ustok, and Harsa (2009)

*$Y_{P/X}$: product yield on cell; $Y_{P/S}$: product yield on substrate; $Y_{X/S}$: cell yield on substrate.
**Medium described by Manera et al. (2008) (g L^{-1}): NH$_4$SO$_4$ (0.8), yeast malt (17.0), KH$_2$PO$_4$ (5.0), MgSO$_4$·7H$_2$O (0.4) and lactose (28.8).
***Medium A used for *L. delbrueckii* ssp. *bulgaricus*: 8% skimmed milk powder and Medium B used for *S. thermophilus*: 5% sweet whey powder, 2% KH$_2$PO$_4$ and 2% peptone.
****MMRS: modified Man Rogosa and Sharpe medium replacing glucose with lactose and MSDB: modified Sabouraud Dextrose broth replacing glucose with lactose.

Table 3 shows the yield factors of β-galactosidase production from different microbial sources using various culture media. It is possible to note that the parameters vary not only according to the microorganism, but also due to the conditions used in the enzymatic production (lactose concentration, aeration and agitation).

Rajoka, Samia, and Sahid (2003) evaluated the production of β-galactosidase by *K. marxianus* and verified that the lactose in the culture medium promoted the highest product yield on cell ($Y_{P/X}$ = 480 U g^{-1}), while other carbon sources, such as arabinose, cellobiose and sucrose, reduced the $Y_{P/X}$. In this study, the culture medium with glucose had the lowest $Y_{P/X}$ (5.0 U g^{-1}).

The calculation of yield factors may also be useful in the optimization of β-galactosidase production. Alves et al. (2010) evaluated the effect of parameters such as agitation and aeration on β-galactosidase activity. The optimal conditions to achieve maximum productivity were found based on an experimental design and using the values of enzymatic activity and yield factors ($Y_{X/S}$ and $Y_{P/X}$) as response variables.

PRODUCTIVITY

The evaluation of the productivity rate is of the primary importance when quantifying production over time. Productivity can also be evaluated in terms of biomass production, mainly when productivity is related to cell growth. The quantity can be expressed as product concentration over fermentation time, where all the microbial growth is considered (Shuler and Kargi 2002).

In processes that focus on enzymatic production, productivity is usually calculated in terms of enzymatic activity over fermentation time. In this case, productivity varies according to the microorganism and yield factors; optimization tools can be used to increase such parameters.

Gomes et al. (2018) evaluated productivity and $Y_{P/S}$ as parameters for the selection of a microorganism as a potential enzyme producer. The lactic acid bacteria *L. reuteri* showed the highest productivity (28.78 U L h^{-1}) and $Y_{P/S}$ (82.32 U g^{-1}) compared to the yeast *K. lactis* and other lactic acid bacteria.

Therefore, the quantification of parameters such as productivity and yield factors are essential to evaluate the efficiency of the fermentative process.

THE DOWNSTREAM PROCESS OF MICROBIAL BETA-GALACTOSIDASE

The downstream process consists of the separation and recovery of enzymes such as β-galactosidase. Due to factors such as the high proportion of water in the culture medium, the presence of organic and inorganic molecules and extra and intracellular metabolites, purification is the most costly stage of enzyme production, representing approximately 80% of the total cost of the process (Goldbeg 1996; Roy and Gupta 2002).

The downstream process can be divided into the following four general steps: clarification (separation of cell debris); concentration or low-resolution purification; high-resolution purification; and finishing and packaging (Desai 2000; Roy and Gupta 2002; Flickinger 2013).

The choice of the separation process varies according to the bioproduct that is formed, which can constitute the same cells inoculated in the process (microbial cells), as well as their intra or extracellular products (Flickinger 2013).

Clarification is one of the downstream steps used to recover biomolecules; solid particles suspended as microbial cells (in their entire form or fragments thereof) are removed, with a consequent reduction in the turbidity of the culture broth. For this purpose, steps such as conventional filtration, centrifugation, membrane separation and/or flocculation could be applied. The choice of type of operation will depend on the size and density of the particles (Desai 2000; Roy and Gupta 2002; Flickinger 2013).

Centrifugation is the most widely used operation to separate the biomass from the supernatant (clarified broth) on an industrial scale. In this process cells are suspended in the culture medium sediment due to acceleration caused by a centrifugal gravitational force (Desai 2000).

When biomolecules are produced intracellularly, such as β-galactosidase from yeasts and bacteria, cellular disruption steps are used. Thus, the additional steps involving separation and purification take longer than those used for extracellular enzymes and are also more expensive (Uzir and Don 2008).

Centrifugation is the most effective process to separate microbial biomass when the particles to be removed are very small, or when the viscosity levels of the fluid is high and the filtration method is not sufficient. Moreover, centrifugation is recommended for recovering β-galactosidase from yeasts and bacteria because the microfiltration procedure might result in obstruction of the membrane pores (Desai 2000; Uzir and Don 2008; Katoh, Horiuchi, and Yoshida 2015).

The low-resolution purification and concentration steps promote the separation of the biomolecules from the culture medium using precipitation, and micro and ultrafiltration (membranes).

Precipitation is one of the most commonly used stages in the initial processes of purification, both in industry and in the laboratory. Salting-out or salting-in can be used for this, with the addition of organic solvents, heat, pH adjustment, polyelectrolytes and metal ions (Lucarini, Kilikian, and Pessoa Junior 2005).

Ultrafiltration is widely used in biotechnological processes during the downstream stage, both for the concentration and purification of proteins and enzymes, mainly for the ultrafiltration of membranes to retain macromolecules in solution (Nobrega, Borges, and Habert 2005).

High-resolution concentration/purification separates molecules based on their physicochemical characteristics (Uzir and Don 2008; Katoh, Horiuchi, and Yoshida 2015). In these steps, chromatographic operations (such as ion exchange chromatography, hydrophobic interaction, molecular exclusion and adsorptive membranes) are applied. The type and density of the charge on the surface of the biomolecules, the specific sites of the protein surface (concentration, adsorption), hydrophobicity, molar mass, and protein adsorption characteristics will all influence the choice of the separation process (Desai 2000), while the low-resolution purification comprises of the separation of biomolecules with similar physicochemical characteristics (Uzir and Don 2008; Katoh, Horiuchi, and Yoshida 2015).

Ion exchange chromatography is one of the most important techniques used for protein purification due to its chemical and physical stability, as well as its exceptional flow characteristics (Silveira et al. 2008). Usually, the column used is Sepharose (formed by 90 µm highly cross-linked 6% agarose beads); it is highly substituted with strong ion exchange groups which remain charged, making it possible to work with a broad range of pH (Silveira et al. 2008; Oliveira de Medeiros, Veiga, and Kalil 2012).

Other techniques can also be used for enzyme purification. β-galactosidase has also been purified from psychotropic *Pseudoalteromonas* sp. isolated from Antarctica. In this study, a high purification yield was reported for a rapid purification scheme using extraction in an aqueous two-phase system followed by hydrophobic interaction chromatography and ultrafiltration techniques (Fernandes et al. 2002).

The efficiency of purification and, consequently, the process yield is affected when ion exchange chromatography is used to separate enzymes in culture media with a high concentration of salts. Multimodal or mixed mode chromatography (MMC) combines two or more types of molecular interaction with simultaneous adsorption phenomena. The purification of β-galactosidase from *K. lactis* was recently performed using MMC with simultaneous ionic and hydrophobic interactions. The purification process resulted in a recovery of 41.0 and 48.2% of total protein concentration and enzymatic activity, respectively (Lima et al. 2016). The use of MMC has been shown to be a promising method for enzyme purification. (Sousa, Prazeres, and Queiroz 2008; Oehme and Peters 2010; Lima et al. 2016).

During the purification steps, other techniques are generally applied to confirm the evolution of the purification process, such as SDS-page electrophoresis, which can verify the feasibility of the method to obtain purified enzymes.

CELL DISRUPTION

Many microbial enzymes are produced intracellularly, and separation techniques are required for enzyme release. Cellular disruption is the first stage and represents one of the most critical steps in the downstream of intracellular biomolecules (Geciova, Bury, and Jelen 2002).

Physical, chemical and enzymatic methods (either combined or not) can be used for microbial cell disruption or for permeabilization to release intracellular products (Harrison 1991; Geciova, Bury, and Jelen 2002).

The bacterial cell wall is composed of several polymers linked to each other, and its composition is different for Gram-positive and Gram-negative bacteria (Ghuysen and Hakenbeck 1994).

Gram-positive bacteria, including lactic acid bacteria (*Lactobacillus*, *Leuconostoc*, *Lactococcus*, *Streptococcus* or *Pediococcus* sp.), *Bacillus* sp. and many others have a lower concentration of lipids when compared to Gram-negative bacteria and, therefore, their resistance to chemical solvents is higher (Salton 1953). Furthermore, a thick layer of peptidoglycan provides high rigidity to the cell wall of Gram-positive bacteria (Salton 1953; Navarre and Schneewind 1999). Gram-negative bacteria have a lower concentration of peptidoglycan and the level of lipopolysaccharides is higher (Takeuchi et al. 1999; Brown et al. 2015).

The yeast cell wall is primarily composed of glucans (branched molecules of glucose units linked by β-1,3 and β-1,6 bonds), mannan (mannose residues linked together by α-1,6 bonds and short side-chains of oligosaccharides) and proteins (mainly enzymes without structural function) (Middelberg 1995). Due to their composition, yeast cell walls are harder to break than bacteria cell walls (Pessoa Junior 2005).

Cell disruption methods are classified as either mechanical (severe) or non-mechanical (gentle methods), which are subdivided into physical, chemical and enzymatic methods. According to Harrison (1991), the technique of cellular disruption, as well as its application, significantly influences the degree of recovery. Factors such as subsequent purification steps (refining), the nature of the processed suspensions, and the form and quality of the final product also influence the quality and degree of purity.

Sonication, French-press, freeze-pressing, grinding and glass-bead homogenization are the most common techniques used in mechanical cell disruption, while enzymatic lysis, which mainly uses lysozyme and solvents (non-mechanical methods), can also be applied.

Table 4. Different cell disruption methods applied to release β-galactosidase from yeast and lactic acid bacteria

Method of cell disruption		Microorganism	Conditions of cell disruption	Reference
Mechanical	Glass beads + vortex stirring	*K. lactis; L. reuteri; L. plantarum, L. johnsonii; L. bulgaricus; L. fermentum; Pediococcus acidilactici; P. pentosaceus*	0.4 g glass beads (4 mm diameter)/mL cell suspension under vortex stirring/15 min	Gomes et al. (2018)
	Glass beads + homogenizer	*S. thermophilus; L. delbrueckii ssp. bulgaricus; L. acidophilus*	The cell suspension was cooled (4°C) and 10 mL glass beads (0.1 mm diameter) were added and homogenized (3,000 rpm)	Ibrahim (2018)
	Bead beating homogenizer + glass beads	*L. plantarum*	1 g of glass beads were added to a cell suspension (1 mL) of 50 mM phosphate buffer and homogenized using a bead beating homogenizer (4,000 – 6,800 rpm)	Geiger et al. (2016)
	Quartz sand + vortex stirring	*L. acidophilus*	Quartz sand (150 µm diameter) was added to cell suspension (0.1 M phosphate buffer) + vortex stirring/10 min	Carevic et al. (2015)
	Ultrasonication	*B. safensis; L. acidophilus*	Cell suspension was sonicated for 300 s (16 KHz at 400 W using a probe of 9.5 mm outer diameter); cell suspension (phosphate buffer) was sonicated (20 KHz, 50 W)/15 min	Nath et al. (2016); Choonia and Lele (2011)
	Ultrasonic bath + glass beads	*K. marxianus*	1.10 g glass beads (0.95 – 1.05 diameter) in ultrasonic bath/40 min	Medeiros et al. (2008)

Table 4. (Continued)

	Method of cell disruption	Microorganism	Conditions of cell disruption	Reference
Non-Mechanical	SDS + chloroform	K. lactis; K. marxianus; S. thermophilus	Cell suspension (0.05 M phosphate buffer) was mixed with 0.9 mL Z buffer* and 100 µL chloroform + 50 µL 0.1% SDS solution/5 min at 37°C; 0.1% SDS solution + 0.1 mL absolute chloroform; 100 µL chloroform + 50 µL 1% SDS solution;	Sangwan et al. (2015) Ornelas et al. (2008); Bansal et al. (2008);
	Isoamyl alcohol	K. lactis	Cell suspension in 0.2 M phosphate buffer with 5 mL isoamyl alcohol and diluted up to 25 mL with the same buffer. The mixture was shaken/15 min (room temperature)	Dagbagli and Goksungur (2008)
	Toluene + ethanol	K. lactis	2% toluene + 10% ethanol with constant shaking	Inchaurrondo, Flores, and Voget (1998)
	Lysozyme	L. delbrueckii ssp. bulgaricus; S. thermophilus	100 mg lysozyme was added to cell suspension. The mixture was incubated at 37°C/15 min. Then, 0.5 mL of 4 M NaCl solution was added and incubated (37°C/50 min)	Tari, Ustok, and Harsa (2009)

*Z buffer: 0.06 M Na_2HPO_4; 0.04 M NaH_2PO_4; 0.01 M KCl and 0.001 M $MgSO_4$

Table 4 shows the different cell disruption methods (mechanical and non-mechanical) commonly used for cell disruption of yeasts and bacteria to release β-galactosidase.

The glass bead technique is one of the most commonly used methods for cell disruption of lactic acid bacteria for enzymatic release in the laboratory (with vortex stirring and beating homogenizer) and on a large-scale (grinding chamber). The cells are disrupted by the shear forces of radial acceleration of glass beads (zirconia and quartz sand can also be used). This process produces heat and efficient cooling of the system is therefore required (Najafpour, Jahanshahi and Najafpour 2007).

Ultrasonication is more effective than other mechanical methods due to the cavitation phenomena, which is a combination of the formation, growth and collapse of vapor bubbles, which are produced by an intense sound wave. In this process, a large amount of sonic energy is converted into mechanical energy (Borthwick et al. 2005; Liu et al. 2013; Gogate 2011; Tangtua 2014).

Different solvents (such as ethanol, toluene, isoamyl alcohol, sodium dodecyl sulfate - SDS - and chloroform) and lysozyme have been reported as non-mechanical methods for releasing β-galactosidase from yeast (*Kluyveromyces* genus) and lactic acid bacteria (*Streptococcus* and *Lactobacillus* genus).

The advantage of using solvents is that they are inexpensive and straightforward; they can also be used prior to mechanical methods, increasing the effectiveness of the cell disruption. However, the toxicity and low specificity of solvents are disadvantages (Berry, Russell, and Stewart 1987; Chaplin and Bucke 1990; Harrison 1991; Roy and Gupta 2002).

Lysozyme is most common bacteriolytic enzyme that is commercially available for large-scale application. The use of enzymatic cell lysis is advantageous due to biological specificity, mild operating conditions, low energy requirements and minimum damage to the product, despite the high cost of the process (Harrison 1991; Tari, Ustok, and Harsa 2009).

Although there are several different methods regarding microbial cell disruption, knowledge of the cell structure and its composition are essential for the choice of an appropriate and effective technique.

After cell disruption and the release of β-galactosidase, separation stages are required, in which the separation of cellular debris (fragments of the cell wall, nucleic acids and cellular organelles) as well as contamination and possible denaturation should be considered (Desai 2000). For this purpose, clarification, followed by centrifugation or membrane separation (micro and ultrafiltration), is required to separate the biomass from the supernatant.

BETA-GALACTOSIDASE ACTIVITY ASSAY

The amount of o-nitrophenol (ONP) released from o-nitrophenol-β-D-galactopyranoside (ONPG) as chromogenic substrate was used to measure β-galactosidase activity based on a modified Miller method (Miller, 1972). The samples were centrifuged, and the absorbance was measured at 420 nm. The nano-moles of ONP that were liberated were determined from a standard curve that measured the change in absorbance produced by various ONP concentrations in the range from 200 to 1,000 nmol. The amount of ONP released/min is directly proportional to the quantity of enzyme, and one unit of β-galactosidase activity (U min^{-1} mL^{-1}) is defined as the amount of enzyme that hydrolyzes 1 nmol of ONPG to o-nitrophenol per minute under standard conditions (Food Chemical Codex 1993).

IMMOBILIZATION OF MICROBIAL BETA-GALACTOSIDASE

Although the enzyme β-galactosidase has numerous applications in the food and dairy industries, its moderate stability is one of the limitations that hinder the general implementation of this biocatalyst at an industrial scale. Thus, there is a need to explore its full potential as a catalyst by adopting suitable strategies for enzyme stabilization. Multimeric enzymes can be stabilized by using proper experimental conditions and genetic tools to cross-link or strengthen the subunit-subunit interaction (Panesar, Kumari, and Panesar 2010). The stability of monomeric or multimeric enzymes can also be enhanced by multipoint and multi-subunit covalent immobilization, as well as enzyme engineering via immobilization.

An immobilized enzyme is defined as "the enzyme physically confined or localized in a certain defined region of space with retention of its catalytic activity, which can be used repeatedly and continuously" (Chibata 1979). β-galactosidase is one of the most studied enzymes in terms of immobilization. Although many studies have described the effective immobilization of β-galactosidase isolated from recombinant *E. coli*, its application in the food industry is complicated because this microorganism is not GRAS (Ladero et al. 2001; Serio et al. 2003).

The advantages of the immobilization process include the following: the immobilized biocatalyst can be reused, decreasing the costs of the process; there is easy separation from the reaction solution compared to free enzymes; there is no contamination of the product by the enzyme; and it provides operational and long-term stability, continuous processing, and multienzyme reaction systems (Dervakos and Webb 1991; Grosová, Rosenberg, and Rebroš 2008).

Techniques and Matrices for the Immobilization of β-Galactosidase

β-Galactosidases have been immobilized by several methods using a variety of matrices. These immobilization methods have included entrapment, cross-linking, adsorption, covalent binding or a combination of the same (Table 5). Since each process has its advantages and drawbacks, the selection of a suitable immobilization method depends on the enzyme, including factors such as the different properties of various β-galactosidases such as molecule weight, protein chain length, and position of the active site (Tanaka and Kawamoto 1999).

Table 5. Different methods for yeasts β-galactosidase immobilization (Table adapted from Grosová, Rosenberg, and Rebroš 2008)

Source of enzyme	Immobilization method	Recovery of activity (%)	References
K. fragilis	- covalent binding on corn grits	8	Siso et al. (1994)
	- covalent binding on cellulose beads	82	Roy and Gupta (2003)
	- covalent binding on porous silanized glass modified by glutaraldehyde	90	Szczodrak (2000)
	- entrapment in alginate-carrageenan gels	-	Mammarella and Rubiolo (2005)
	- adsorption on phenol-formaldehyde resin	23	Woudenberg-van Oosterom et al. (1998)
	- adsorption onto bone powder	83	Carpio et al. (2000)
K. lactis	- covalent binding onto glutaraldehyde-agarose	36 - 40	Giacomini et al. (2001)
	- covalent binding onto thiolsulfinate-agarose	60	
	- covalent binding on graphite surface	0.01	Zhou and Dong Chen (2001)
K. marxianus	- covalent binding on oxides supports: alumina, silica, silicated alumina	< 5	Serio et al. (2003)

Covalent Binding

This method is mostly used for β-galactosidase immobilization (Table 5). A covalent bond links enzyme to the support through the functional groups in the enzymes that are not essential for catalytic activity. Compared to other techniques this method has the following advantages: the enzymes do not leak or detach from the carrier; and the biocatalyst can easily interact with the substrate once it is attached to the surface. Several matrices have been used to immobilize β-galactosidase. Oxide materials, such as alumina, silica and silicated alumina were used for the covalent binding of β-galactosidase from *K. marxianus* and applied in lactose hydrolysis processes. Despite the fact that the immobilizate showed good stability, the immobilization yields were less than 5% (Serio et al. 2003). β-galactosidase from *K. fragilis* was covalently linked to silanized porous glass beads via amino groups, using glutaraldehyde. The coupling efficiency was very high; more than 90% of the enzyme was active and 87.5% of the protein was bound to the support.

Immobilization increased the enzymatic thermal stability and shifted the optimal pH to a more alkaline level (7.7) compared to the free enzyme (6.6). Zhou and Chen (2001) observed a 25-fold increase in the thermal stability of proteins after immobilization: the immobilized enzymes had a half-life of 50 days at 50°C and more than one year at 40°C. Giacomini et al. (1998) compared the properties of immobilized *K. lactis* β-galactosidase using two different coupling carriers: glutaraldehyde-agarose gel and thiosulfinate-agarose gel. Glutaraldehyde agarose exhibited a lower yield after immobilization (36 – 40%) than thiosulfinate-agarose (60 – 85%), but better thermal properties.

Physical Adsorption

Physical adsorption is the simplest and the oldest method of immobilizing enzymes onto carriers. Immobilization by adsorption is based on the physical interactions between the biocatalyst and the carrier, such as hydrogen bonding, hydrophobic interactions, van der Waals force, and combinations of the same. Despite its simplicity, this immobilization method is significantly limited by the tendency of the enzyme to desorb from the support, as well as sensitivity to environmental conditions such as temperature and the concentration of ions (Tanaka and Kawamoto 1999). Immobilized β-galactosidase particles (size 1 – 2 mm) that were prepared by physical adsorption on porous ceramic support and intermolecular cross-linking with glutaraldehyde reached a binding efficiency of 80% and good operational stability.

Taqieddin and Amiji (2004) developed a new encapsulation method in which alginate-chitosan core-shell microcapsules were formed for the immobilization of β-galactosidase. The enzymes were localized and protected in the inner biocompatible alginate core, while the outer chitosan shell dictated the transport properties. When using Ca^{2+} ions for crosslinking alginate, microcapsules with a liquid core were produced with 60% loading efficiency. When using Ba^{2+} ions, microcapsules with a solid core were produced and 100% loading efficiency was obtained.

Many of the carriers used for the immobilization of β-galactosidase applied in GOS production have been different types of microparticles, such as ion-exchange resins (Matsumoto et al. 1989), chitosan beads (Sheu et al. 1998; Shin, Park, and Yang 1998), cellulose beads (Kmínková, Prošková, and Kučera 1988) and agarose beads (Berger, Lee, and Lacroix 1995). The immobilized enzyme in these particle carriers often resulted in a 20 – 30% reduction in GOS yield due to the introduction of mass transfer resistance into the system (Sheu et al. 1998; Shin, Park, and Yang 1998). An appropriate method of β-galactosidase immobilization leading to increasing transgalactosylation activity is still to be developed.

Gaur et al. (2006) compared the following two different techniques for *A. oryzae* β-galactosidase immobilization in terms of stability and efficiency in GOS synthesis: covalent coupling to chitosan (beads form); and aggregation by cross-linking (using glutaraldehyde). When 20% (w/v) of lactose was used, the chitosan-immobilized β-galactosidase gave maximum trisaccharide yield (17.3% of the total sugar) within 2 h, compared to 10% obtained with free enzymes, and 4.6 obtained with cross-linked aggregates.

The immobilization of β-galactosidases can dramatically affect the enzyme's properties; e.g., pH and temperature stability, kinetic parameters, etc. (Rossi et al. 1999; Sun et al. 1999; Ladero, Santos, and García-Ochoa 2000). If an adequate technique is applied, immobilization can improve the properties of β-galactosidases such as the stability of the enzymes at high or low pH and temperatures.

The immobilization of β-galactosidase from *Aspergillus oryzae* on cotton cloth activated with *p*-toluenesulfonyl chloride (tosyl chloride) was studied by Albayrak and Yang (2002). Of the different fibrous matrices that were tested knitted cotton cloth showed the highest immobilized enzyme activity. Approximately 50 mg of the enzyme by gram of cotton cloth was immobilized, resulting in a protein coupling efficiency of approximately 85% and enzyme activity yield of approximately 55%. The immobilized enzymes had a half-life of about 50 days at 50°C and more than one year at 40 C, an improvement of 25 – 28-fold compared to free enzymes.

Alonso et al. (2005) investigated the use of glutaraldehyde chemistry to stabilize glutaryl acylase (GAC) immobilized on different aminated supports (ethylendiamine (EA) or polyethylenimine (PEI) coated supports), as well as the effect of glutaraldehyde on both stability and activity. It was determined that the immobilization on aminated supports increased the enzymatic stability and that this stabilization increased with the size of the polyethylenimine. The treatment with glutaraldehyde presented a low impact on enzymatic activity (activity recoveries were more than 80%) and significantly improved enzymatic stability. A similar procedure using enzymes immobilized on supports that could not react with glutaraldehyde did not give rise to stabilization, suggesting that this stabilization was due to a reaction between the protein and the support through glutaraldehyde chemistry.

A. oryzae β-galactosidase was immobilized on an inexpensive bioaffinity support (concanavalin A-cellulose) in which the adsorbed and cross-linked β-galactosidase preparation maintained 78% of the initial activity. The optimum temperature was increased from 50 to 60°C for the immobilized β-galactosidase. The cross-linked adsorbed enzyme maintained 93% activity after one month of storage, while the native enzyme showed only 63% activity under similar incubation conditions (Ansari and Husain 2010).

A kinetic model for lactose using immobilized β-galactosidase from *K. fragilis* has also been developed. The immobilized enzymes were active at a low temperature (5°C); this method could also be applied to freeze dairy products to avoid lactose crystallization and to enhance the digestibility and flavor of such products (Ladero, Santos, and García-Ochoa 2000).

Industrial Applications of Immobilized Beta-Galactosidase

Numerous immobilization systems for lactose hydrolysis have been investigated; however, only a few of them were successfully scaled up, and even fewer have been applied at an industrial or pilot scale. This is mainly because the materials and methods used for enzyme immobilization are either too expensive or difficult to use on an industrial scale (Albayrak and Yang 2002). The first industrial application of immobilized β-galactosidase in the food industry was performed by SnamProgetti (Italy) and Sumitomo Chemicals (Japan) in the 1970s (Gekas and Lopez-Leiva 1985). Since 1977, the Valio laboratory in Finland has used fungal β-galactosidase adsorbed to phenol-formaldehyde resin (Duolite ES-762) for whey processing. In this process, whey and whey permeate are hydrolyzed continuously by pumping through the column (Marconi and Morisi 1979; Gekas and Lopez-Leiva 1985). Thus, immobilization technology plays a significant role in milk and whey processing. The use of immobilized β-galactosidase in the hydrolysis of lactose is a topic of considerable scientific and technological interest.

CONCLUSION

β-galactosidase is widely used in the food processing industry and has several nutritional and environmental applications. This enzyme is found in plants, animals and microorganisms. Although only the enzymes produced by fungi and yeasts are applied industrially, the use of lactic acid bacteria as a source of β-galactosidase has attracted considerable interest due to its GRAS status and its nutritional and technological benefits. Different immobilization techniques have been studied in order to increase the range of applications and this has generated highly efficient processes that have provided excellent yields. Although β-galactosidase has already been studied to a large degree there is still growing research regarding issues such as its applications, its production by different microorganisms using agroindustrial by-products, different purification methods and immobilization techniques. All of these factors make it one of the most important enzymes for technological applications.

REFERENCES

Abbasalizadeh, Saeed, Mohammad Amin Hejazi, and Mahdi Pesaran Hajiabbas. 2015. "Kinetics of β-Galactosidase Production by Lactobacillus Bulgaricus During PH Controlled Batch Fermentation in Three Commercial Bulk Starter Media." *Applied Food Biotechnology* 2 (4): 39–47. doi:10.22037/afb.v2i4.9512.

Agrawal, Sanjeev, H.M. Sonawat, and S.M. Dutta. 1982. "Thermostable Beta-Galactosidase from Fungi." *Journal of Dairy Science* 65 (5): 866–70. doi:10.3168/JDS.S0022-0302(82)82278-9.

Akcan, Nurullah. 2011. "High Level Production of Extracellular β-Galactosidase from Bacillus Licheniformis ATCC 12759 in Submerged Fermentation." *African Journal of Microbiology Research* 5 (26): 4615–21. doi:10.5897/AJMR11.716.

Al- jazairi, M., S. Abou-ghorra, Y. Bakri, and M. Mustafa. 2015. "Optimization of β-Galactosidase Production by Response Surface Methodology Using Locally Isolated Kluyveromyces Marxianus." *International Food Research Journal* 22 (4): 1361–67. http://www.ifrj.upm.edu.my/22 (04) 2015/(7).pdf.

Albayrak, Nedim, and Shang-Tian Yang. 2002. "Immobilization of Aspergillus Oryzae β-Galactosidase on Tosylated Cotton Cloth." *Enzyme and Microbial Technology* 31 (4): 371–83. doi:10.1016/S0141-0229(02)00115-1.

Alikkunju, Aneesa P, Neethu Sainjan, Reshma Silvester, Ajith Joseph, Mujeeb Rahiman, Ally C Antony, Radhakrishnan C Kumaran, and Mohamed Hatha. 2016. "Screening and Characterization of Cold-Active β-Galactosidase Producing Psychrotrophic Enterobacter Ludwigii from the Sediments of Arctic Fjord." *Applied Biochemistry and Biotechnology* 180 (3): 477–90. doi:10.1007/s12010-016-2111-y.

Alonso, Noelia, Fernando López-Gallego, Lorena Betancor, Aurelio Hidalgo, Cesar Mateo, Jose M. Guisan, and Roberto Fernandez-Lafuente. 2005. "Immobilization and Stabilization of Glutaryl Acylase on Aminated Sepabeads Supports by the Glutaraldehyde Crosslinking

Method." *Journal of Molecular Catalysis B: Enzymatic* 35 (1): 57–61. doi:10.1016/j.molcatb.2005.05.007.

Alves, Fernanda Germano, Francisco Maugeri Filho, Janaína Fernandes de Medeiros Burkert, and Susana Juliano Kalil. 2010. "Maximization of β-Galactosidase Production: A Simultaneous Investigation of Agitation and Aeration Effects." *Applied Biochemistry and Biotechnology* 160 (5): 1528–39. doi:10.1007/s12010-009-8683-z.

Ansari, Shakeel Ahmed, and Qayyum Husain. 2010. "Lactose Hydrolysis by β Galactosidase Immobilized on Concanavalin A-Cellulose in Batch and Continuous Mode." *Journal of Molecular Catalysis B: Enzymatic* 63 (1): 68–74. doi:10.1016/j.molcatb.2009.12.010.

———. 2012. "Lactose Hydrolysis from Milk/Whey in Batch and Continuous Processes by Concanavalin A-Celite 545 Immobilized Aspergillus Oryzae β Galactosidase." *Food and Bioproducts Processing* 90 (2): 351–59. doi:10.1016/J.FBP.2011.07.003.

Ansari, Shakeel Ahmed, and Rukhsana Satar. 2012. "Recombinant β-Galactosidases – Past, Present and Future: A Mini Review." *Journal of Molecular Catalysis B: Enzymatic* 81 (September): 1–6. doi:10.1016/J.MOLCATB.2012.04.012.

Arukha, Ananta Prasad, Bidhan Chandra Mukhopadhyay, Suranjita Mitra, and Swadesh Ranjan Biswas. 2015. "A Constitutive Unregulated Expression of β-Galactosidase in Lactobacillus Fermentum M1." *Current Microbiology* 70 (2): 253–59. doi:10.1007/s00284-014-0711-8.

Bansal, Sunil, Harinder Singh Oberoi, Gurpreet Singh Dhillon, and R.T. Patil. 2008. "Production of β-Galactosidase by Kluyveromyces Marxianus MTCC 1388 Using Whey and Effect of Four Different Methods of Enzyme Extraction on β-Galactosidase Activity." *Indian Journal of Microbiology* 48 (3): 337–41. doi:10.1007/s12088-008-0019-0.

Benjamins, Eric, Laura Boxem, Janke KleinJan-Noeverman, and Ton A. Broekhuis. 2014. "Assessment of Repetitive Batch-Wise Synthesis of Galacto-Oligosaccharides from Lactose Slurry Using Immobilised β-

Galactosidase from *Bacillus Circulans*." *International Dairy Journal* 38 (2): 160–68. doi:10.1016/j.idairyj.2014.03.011.

Berger, J.L., B.H. Lee, and C. Lacroix. 1995. "Immobilization of ?-Galactosidases from Thermus Aquaticus YT-1 for Oligosaccharides Synthesis." *Biotechnology Techniques* 9 (8): 601–6. doi:10.1007/BF00152452.

Berry, D.R., I. Russell, and G.G. Stewart, eds. 1987. *Yeast Biotechnology*. Dordrecht: Springer Netherlands. doi:10.1007/978-94-009-3119-0.

Borthwick, K.A.J., W.T. Coakley, M.B. McDonnell, H. Nowotny, E. Benes, and M. Gröschl. 2005. "Development of a Novel Compact Sonicator for Cell Disruption." *Journal of Microbiological Methods* 60 (2): 207–16. doi:10.1016/j.mimet.2004.09.012.

Brown, Lisa, Julie M. Wolf, Rafael Prados-Rosales, and Arturo Casadevall. 2015. "Through the Wall: Extracellular Vesicles in Gram-Positive Bacteria, Mycobacteria and Fungi." *Nature Reviews. Microbiology* 13 (10): 620–30. doi:10.1038/nrmicro3480.

Campuzano, Susana, Beatriz Serra, Daniel Llull, José L García, and Pedro García. 2009. "Cloning, Expression, and Characterization of a Peculiar Choline-Binding -Galactosidase from *Streptococcus Mitis*." *Applied and Environmental Microbiology* 75 (18): 5972–80. doi:10.1128/AEM.00618-09.

Cardoso, Beatriz B, Sara C Silvério, Luís Abrunhosa, José A Teixeira, and Lígia R Rodrigues. 2017. "β-Galactosidase from *Aspergillus Lacticoffeatus*: A Promising Biocatalyst for the Synthesis of Novel Prebiotics." *International Journal of Food Microbiology* 257 (September): 67–74. doi:10.1016/J.IJFOODMICRO.2017.06.013.

Carevic, Milica, Maja Vukasinovic-Sekulic, Sanja Grbavcic, Marija Stojanovic, Mladen Mihailovic, Aleksandra Dimitrijevic, and Dejan Bezbradica. 2015. "Optimization of β-Galactosidase Production from Lactic Acid Bacteria." *Hemijska Industrija* 69 (3): 305–12. doi:10.2298/HEMIND140303044C.

Carpio, C., P. González, J. Ruales, and F. Batista-Viera. 2000. "Bone-Bound Enzymes for Food Industry Application." *Food Chemistry* 68 (4): 403–9. doi:10.1016/S0308-8146(99)00193-4.

Chaplin, Martin F., and C. Bucke. 1990. *Enzyme Technology*. New York: Cambridge University Press.

Chibata, I., ed. 1979. *Immobilized Enzymes, Research and Development*. Tokyo: John Wiley & Sons Inc. Publication.

Choonia, Huzaifa S., and S.S. Lele. 2011. "β-Galactosidase Release Kinetics during Ultrasonic Disruption of Lactobacillus Acidophilus Isolated from Fermented Eleusine Coracana." *Food and Bioproducts Processing* 89 (4): 288–93. doi:10.1016/j.fbp.2010.08.009.

Clark, Douglas S., and Harvey W. Blanch. 1997. *Biochemical Engineering*. 2nd ed. CRC Press.

Couri, S, and M.C.T. Damaso. 2015. *Knowledge Tree: Food Technology*. 2015. http://www.agencia.cnptia.embrapa.br/gestor/tecnologia_de_alimentos/arvore/CONT000fid5sgif02wyiv80z4s473v6o7sud.html.

Dagbagli, Seval, and Yekta Goksungur. 2008. "Optimization of Beta-Galactosidase Production Using Kluyveromyces Lactis NRRL Y-8279 by Response Surface Methodology." *Electronic Journal of Biotechnology* 44 (4): 1–12. doi:10.4067/S0717-34582008000400011.

Damodaran, Srinivasan., Kirk L Parkin, and Owen R Fennema. 2008. *Fennema's Food Chemistry*. 4 th. Boca Raton: CRC Press/Taylor & Francis.

Demirhan, Elçin, Dilek Kılıç Apar, and Belma Özbek. 2010. "A Modelling Study on Hydrolysis of Whey Lactose and Stability of β-Galactosidase." *Korean Journal of Chemical Engineering* 27 (2): 536–45. doi:10.1007/s11814-010-0062-5.

Dervakos, George A., and Colin Webb. 1991. "On the Merits of Viable-Cell Immobilisation." *Biotechnology Advances*. Elsevier. doi:10.1016/0734-9750(91)90733-C.

Desai, Mohamed A., ed. 2000. *Downstream Processing of Proteins*. Vol. 9. Methods in Biotechnology. Totowa, NJ: Humana Press. doi:10.1007/978-1-59259-027-8.

Dominguez-Jimenez, Jose Luis, Antonio Fernandez-Suarez, Sara Ruiz-Tajuelos, Juan Jesus Puente-Gutierrez, and Antonio Cerezo-Ruiz. 2014. "Lactose Tolerance Test Shortened to 30 Minutes: An Exploratory Study of Its Feasibility and Impact." *Revista Espanola de*

Enfermedades Digestivas : Organo Oficial de La Sociedad Espanola de Patologia Digestiva 106 (6): 381–85. http://scielo.isciii.es/pdf/diges/v106n6/original2.pdf.

Dragosits, Martin, Stefan Pflügl, Simone Kurz, Ebrahim Razzazi-Fazeli, Iain B.H. Wilson, and Dubravko Rendic. 2014. "Recombinant Aspergillus β-Galactosidases as a Robust Glycomic and Biotechnological Tool." *Applied Microbiology and Biotechnology* 98 (8): 3553–67. doi:10.1007/s00253-013-5192-3.

Economou, Ch.N., A. Makri, G. Aggelis, S. Pavlou, and D.V. Vayenas. 2010. "Semi-Solid State Fermentation of Sweet Sorghum for the Biotechnological Production of Single Cell Oil." *Bioresource Technology* 101 (4): 1385–88. doi:10.1016/j.biortech.2009.09.028.

Fernandes, S., B. Geueke, O. Delgado, J. Coleman, and R. Hatti-Kaul. 2002. "Beta-Galactosidase from a Cold-Adapted Bacterium: Purification, Characterization and Application for Lactose Hydrolysis." *Applied Microbiology and Biotechnology* 58 (3): 313–21. doi:10.1007/s00253-001-0905-4.

Fischer, Christin, and Thomas Kleinschmidt. 2015. "Synthesis of Galactooligosaccharides Using Sweet and Acid Whey as a Substrate." *International Dairy Journal* 48: 15–22. doi:10.1016/J.IDAIRYJ.2015.01.003.

Fischer, Janaína, Carla Zanella Guidini, Larissa N. Soares Santana, Miriam Maria de Resende, Vicelma Luiz Cardoso, and Eloízio Júlio Ribeiro. 2013. "Optimization and Modeling of Lactose Hydrolysis in a Packed Bed System Using Immobilized β-Galactosidase from Aspergillus Oryzae." *Journal of Molecular Catalysis B: Enzymatic* 85–86: 178–86. doi:10.1016/j.molcatb.2012.09.008.

Flickinger, Michael C., ed. 2013. *Downstream Industrial Biotechnology: Recovery and Purification*. Hoboken: John Wiley & Sons Inc. Publication.

Food and Agricultural Organization. 2010. *Status of and Prospects for Smallholder Milk Production – A Global Perspective*. Edited by Torsten Hemme and Joachim Otte. Rome: Food and Agricultural Organization. http://www.fao.org/docrep/012/i1522e/i1522e00.pdf.

———. 2013. *Milk and Dairy Products in Human Nutrition*. Edited by Ellen Muehlhoff, Anthony Bennett, and Deirdre McMahon. Rome: Food and Agricultural Organization. http://www.fao.org/docrep/018/i3396e/i3396e.pdf.

Food Chemical Codex. 1993. "Lactase (β-Galactosidase) Activity." In *Food Chemical Codex*, 3rd ed., 491. Washington, DC: National Academy Press.

Fowler, A.V., and I. Zabin. 1978. "Amino Acid Sequence of Beta-Galactosidase. XI. Peptide Ordering Procedures and the Complete Sequence." *The Journal of Biological Chemistry* 253 (15): 5521–25. http://www.ncbi.nlm.nih.gov/pubmed/97298.

Freitas, Fernanda F., Líbia D.S. Marquez, Gustavo P. Ribeiro, Gabriela C. Brandão, Vicelma L. Cardoso, and Eloízio J. Ribeiro. 2011. "A Comparison of the Kinetic Properties of Free and Immobilized Aspergillus Oryzae β-Galactosidase." *Biochemical Engineering Journal* 58–59: 33–38. doi:10.1016/j.bej.2011.08.011.

Gänzle, Michael G., Gottfried Haase, and Paul Jelen. 2008. "Lactose: Crystallization, Hydrolysis and Value-Added Derivatives." *International Dairy Journal* 18 (7): 685–94. doi:10.1016/j.idairyj.2008.03.003.

Gaur, Ruchi, Hema Pant, Ruchi Jain, and S.K. Khare. 2006. "Galacto-Oligosaccharide Synthesis by Immobilized Aspergillus Oryzae β-Galactosidase." *Food Chemistry* 97 (3): 426–30. doi:10.1016/J.FOODCHEM.2005.05.020.

Geciova, Jana, Dean Bury, and Paul Jelen. 2002. "Methods for Disruption of Microbial Cells for Potential Use in the Dairy Industry—a Review." *International Dairy Journal* 12 (6): 541–53. doi:10.1016/S0958-6946(02)00038-9.

Geiger, Barbara, Hoang-Minh Nguyen, Stefanie Wenig, Hoang Anh Nguyen, Cindy Lorenz, Roman Kittl, Geir Mathiesen, Vincent G.H. Eijsink, Dietmar Haltrich, and Thu-Ha Nguyen. 2016. "From By-Product to Valuable Components: Efficient Enzymatic Conversion of Lactose in Whey Using β-Galactosidase from *Streptococcus*

Thermophilus." *Biochemical Engineering Journal* 116: 45–53. doi:10.1016/j.bej.2016.04.003.

Gekas, V., and M. Lopez-Leiva. 1985. *Hydrolysis of Lactose: A Literature Review.* 20 (1): 2–12.

Ghuysen, J.M., and R. Hakenbeck, eds. 1994. *Bacterial Cell Wall.* 1st ed. Elsevier Science.

Giacomini, Cecilia, Gabriela Irazoqui, Francisco Batista-Viera, and Beatriz M. Brena. 2001. "Influence of the Immobilization Chemistry on the Properties of Immobilized β-Galactosidases." *Journal of Molecular Catalysis B: Enzymatic* 11 (4–6): 597–606. doi:10.1016/S1381-1177(00)00056-4.

Giacomini, Cecilia, Andrea Villarino, Laura Franco-Fraguas, and Francisco Batista-Viera. 1998. "Immobilization of β-Galactosidase from *Kluyveromyces Lactis* on Silica and Agarose: Comparison of Different Methods." *Journal of Molecular Catalysis B: Enzymatic* 4 (5–6): 313–27. doi:10.1016/S1381-1177(98)00071-X.

Gibson, Glenn R., Hollie M. Probert, Jan Van Loo, Robert A. Rastall, and Marcel B. Roberfroid. 2004. "Dietary Modulation of the Human Colonic Microbiota: Updating the Concept of Prebiotics." *Nutrition Research Reviews* 17 (02): 259. doi:10.1079/NRR200479.

Gogate, Parag R. 2011. "Hydrodynamic Cavitation for Food and Water Processing." *Food and Bioprocess Technology* 4 (6): 996–1011. doi:10.1007/s11947-010-0418-1.

Goldbeg, E. 1996. *Handbook of Downstream Processing.* Edited by Elliott Goldberg. 1st ed. Dordrecht: Springer Netherlands. doi:10.1007/978-94-009-1563-3.

Gomes, Tatiane A., Luiza B. Santos, Alessandro Nogueira, and Michele R. Spier. 2018. "Increase in an Intracellular β-Galactosidase Biosynthesis Using L. Reuteri NRRL B-14171, Inducers and Alternative Low-Cost Nitrogen Sources under Submerged Cultivation." *International Journal of Food Engineering* 14 (3): 1–12. doi:10.1515/ijfe-2017-0333.

Grosová, Zuzana, Michal Rosenberg, and Martin Rebroš. 2008. "Perspectives and Applications of Immobilised β-Galactosidase in

Food Industry – a Review." *Czech Journal of Food Sciences* 26 (1): 1–14. doi:10.17221/1134-CJFS.

Gruber, Helen E., Jane A. Ingram, H. James Norton, and Edward N. Hanley. 2007. "Senescence in Cells of the Aging and Degenerating Intervertebral Disc." *Spine* 32 (3): 321–27. doi:10.1097/01.brs.0000253960.57051.de.

Guven, Reyhan Gul, Alevcan Kaplan, Kemal Guven, Fatma Matpan, and Mehmet Dogru. 2011. "Effects of Various Inhibitors on β-Galactosidase Purified from the T*hermoacidophilic Alicyclobacillus Acidocaldarius* Subsp. *Rittmannii* Isolated from Antarctica." *Biotechnology and Bioprocess Engineering* 16 (1): 114–19. doi:10.1007/s12257-010-0070-7.

Haider, Toshiba, and Qayyum Husain. 2007. "Calcium Alginate Entrapped Preparations of Aspergillus Oryzae β Galactosidase: Its Stability and Applications in the Hydrolysis of Lactose." *International Journal of Biological Macromolecules* 41 (1): 72–80. doi:10.1016/j.ijbiomac.2007.01.001.

———. 2009. "Immobilization of β-Galactosidase by Bioaffinity Adsorption on Concanavalin A Layered Calcium Alginate–starch Hybrid Beads for the Hydrolysis of Lactose from Whey/Milk." *International Dairy Journal* 19 (3): 172–77. doi:10.1016/j.idairyj.2008.10.005.

Hansen, Gustav H., Mette Lübeck, Jens C. Frisvad, Peter S. Lübeck, and Birgitte Andersen. 2015. "Production of Cellulolytic Enzymes from Ascomycetes: Comparison of Solid State and Submerged Fermentation." *Process Biochemistry* 50 (9): 1327–41. doi:10.1016/j.procbio.2015.05.017.

Harju, M., H. Kallioinen, and O. Tossavainen. 2012. "Lactose Hydrolysis and Other Conversions in Dairy Products: Technological Aspects." *International Dairy Journal* 22 (2): 104–9. doi:10.1016/J.IDAIRYJ.2011.09.011.

Harrison, Susan T.L. 1991. "Bacterial Cell Disruption: A Key Unit Operation in the Recovery of Intracellular Products." *Biotechnology Advances* 9 (2): 217–40. doi:10.1016/0734-9750(91)90005-G.

Heyman, M.B. 2006. "Lactose Intolerance in Infants, Children, and Adolescents." *Pediatrics* 118 (3): 1279–86. doi:10.1542/peds.2006-1721.

Hidalgo-Morales, Madeleine, Victor Robles-Olvera, and Hugo S Garcia. 2005. "Lactobacillus Reuteri Beta-Galactosidase Activity and Low Milk Acidification Ability." *Canadian Journal of Microbiology* 51 (3): 261–67. doi:10.1139/w04-134.

Holsinger, V.H. 1997. "Physical and Chemical Properties of Lactose." In *Advanced Dairy Chemistry Volume 3*, edited by P.F. Fox, 1–38. Boston, MA: Springer US. doi:10.1007/978-1-4757-4409-5_1.

Hsu, C.A., R.C. Yu, and C.C. Chou. 2005. "Production of β-Galactosidase by Bifidobacteria as Influenced by Various Culture Conditions." *International Journal of Food Microbiology* 104 (2): 197–206. doi:10.1016/j.ijfoodmicro.2005.02.010.

Huber, R.E., and M.T. Gaunt. 1983. "Importance of Hydroxyls at Positions 3, 4, and 6 for Binding to the 'Galactose' Site of Beta-Galactosidase (*Escherichia Coli*)." *Archives of Biochemistry and Biophysics* 220 (1): 263–71. http://www.ncbi.nlm.nih.gov/pubmed/6402986.

Husain, Qayyum. 2010. "β Galactosidases and Their Potential Applications: A Review." *Critical Reviews in Biotechnology* 30 (1): 41–62. doi:10.3109/07388550903330497.

Ibrahim, Alaa H. 2018. "Enhancement of β-Galactosidase Activity of Lactic Acid Bacteria in Fermented Camel Milk." *Emirates Journal of Food and Agriculture* 30 (4): 256. doi:10.9755/ejfa.2018.v30.i4.1660.

Inchaurrondo, V.A., M.V. Flores, and C.E. Voget. 1998. "Growth and β-Galactosidase Synthesis in Aerobic Chemostat Cultures of *Kluyveromyces Lactis*." *Journal of Industrial Microbiology and Biotechnology* 20 (5): 291–98. doi:10.1038/sj.jim.2900526.

Isobe, Kimiyasu, Naomi Takahashi, Serina Chiba, Miho Yamashita, and Takahumi Koyama. 2013. "Acidophilic Fungus, Teratosphaeria Acidotherma AIU BGA-1, Produces Multiple Forms of Intracellular β-Galactosidase." *Journal of Bioscience and Bioengineering* 116 (2): 171–74. doi:10.1016/j.jbiosc.2013.02.018.

Jacobson, R.H., X.J. Zhang, R.F. DuBose, and B.W. Matthews. 1994. "Three-Dimensional Structure of Beta-Galactosidase from E. Coli." *Nature* 369 (6483): 761–66. doi:10.1038/369761a0.

Jellema, P., F.G. Schellevis, D.A.W.M. van der Windt, C.M.F. Kneepkens, and H.E. van der Horst. 2010. "Lactose Malabsorption and Intolerance: A Systematic Review on the Diagnostic Value of Gastrointestinal Symptoms and Self-Reported Milk Intolerance." *QJM: An International Journal of Medicine* 103 (8): 555–72. doi:10.1093/qjmed/hcq082.

Juers, Douglas H., Brian W. Matthews, and Reuben E. Huber. 2012. "LacZ β-Galactosidase: Structure and Function of an Enzyme of Historical and Molecular Biological Importance." *Protein Science* 21 (12): 1792–1807. doi:10.1002/pro.2165.

Kang, I.K., S. G Suh, K.C. Gross, and J.K. Byun. 1994. "N-Terminal Amino Acid Sequence of Persimmon Fruit Beta-Galactosidase." *Plant Physiology* 105 (3): 975–79. https://www.ncbi.nlm.nih.gov/pubmed/8058842.

Katoh, Shigeo, Jun-ichi Horiuchi, and Fumitake Yoshida. 2015. *Biochemical Engineering: A Textbook for Engineers, Chemists and Biologists.* 2nd ed. New Jersey: Wiley-VCH.

Klein, Manuela P., Lucas P. Fallavena, Jéssie da N. Schöffer, Marco A.Z. Ayub, Rafael C. Rodrigues, Jorge L. Ninow, and Plinho F. Hertz. 2013. "High Stability of Immobilized β-d-Galactosidase for Lactose Hydrolysis and Galactooligo saccharides Synthesis." *Carbohydrate Polymers* 95 (1): 465–70. doi:10.1016/j.carbpol.2013.02.044.

Kmínková, Milena, Alexandra Prošková, and Jiří Kučera. 1988. "Immobilization of Mold Beta-Galactosidase." *Collection of Czechoslovak Chemical Communications* 53 (12): 3214–19. doi:10.1135/cccc19883214.

Ladero, M., A. Santos, and F. García-Ochoa. 2000. "Kinetic Modeling of Lactose Hydrolysis with an Immobilized β-Galactosidase from *Kluyveromyces Fragilis*." *Enzyme and Microbial Technology* 27 (8): 583–92. doi:10.1016/S0141-0229(00)00244-1.

Ladero, M., A. Santos, J.L. García, and F. García-Ochoa. 2001. "Activity over Lactose and ONPG of a Genetically Engineered β-Galactosidase from Escherichia Coli in Solution and Immobilized: Kinetic Modelling." *Enzyme and Microbial Technology* 29 (2–3): 181–93. doi:10.1016/S0141-0229(01)00366-0.

Levitt, Michael, Timothy Wilt, and Aasma Shaukat. 2013. "Clinical Implications of Lactose Malabsorption Versus Lactose Intolerance." *Journal of Clinical Gastroenterology* 47 (6): 471–80. doi:10.1097/MCG.0b013e3182889f0f.

Li, Jingjie, Wen Zhang, Chuan Wang, Qian Yu, Ruirui Dai, and Xiaofang Pei. 2012. "Lactococcus Lactis Expressing Food-Grade β-Galactosidase Alleviates Lactose Intolerance Symptoms in Post-Weaning Balb/c Mice." *Applied Microbiology and Biotechnology* 96 (6): 1499–1506. doi:10.1007/s00253-012-3977-4.

Lima, Micael de Andrade, Maria de Fátima Matos de Freitas, Luciana Rocha Barros Gonçalves, and Ivanildo José da Silva Junior. 2016. "Recovery and Purification of a *Kluyveromyces Lactis* β-Galactosidase by Mixed Mode Chromatography." *Journal of Chromatography B* 1015–1016: 181–91. doi:10.1016/J.JCHROMB.2016.01.053.

Liu, Dan, Xin-An Zeng, Da-Wen Sun, and Zhong Han. 2013. "Disruption and Protein Release by Ultrasonication of Yeast Cells." *Innovative Food Science & Emerging Technologies* 18: 132–37. doi:10.1016/j.ifset.2013.02.006.

Liu, G.X., J. Kong, W.W. Lu, W.T. Kong, H. Tian, X.Y. Tian, and G.C. Huo. 2011. "β-Galactosidase with Transgalactosylation Activity from *Lactobacillus Fermentum* K4." *Journal of Dairy Science* 94 (12): 5811–20. doi:10.3168/jds.2011-4479.

Liu, Zongbao, Chao Zhao, Yunjin Deng, Yifan Huang, and Bin Liu. 2015. "Characterization of a Thermostable Recombinant β-Galactosidase from a Thermophilic Anaerobic Bacterial Consortium YTY-70." *Biotechnology & Biotechnological Equipment* 29 (3): 547–54. doi:10.1080/13102818.2015.1015244.

López Leiva, Miguel H., and Mónica Guzman. 1995. "Formation of Oligosaccharides during Enzymic Hydrolysis of Milk Whey

Permeates." *Process Biochemistry* 30 (8): 757–62. doi:10.1016/0032-9592(95)00006-2.

Lucarini, A.C., B.V. Kilikian, and A. Pessoa Junior. 2005. "Precipitation." In *Purification of Biotechnological Productsotechnological Products*, 1st ed., 89–113. São Paulo: Manole.

Machado, Juliana Ribeiro, Marina Born Behling, Anna Rafaela Cavalcante Braga, and Susana Juliano Kalil. 2015. "β-Galactosidase Production Using Glycerol and Byproducts: Whey and Residual Glycerin." *Biocatalysis and Biotransformation* 33 (4): 208–15. doi:10.3109/10242422.2015.1100363.

Macris, B.J., and P Markakis. 1981. "Characterization of Extracellular Beta-d-Galactosidase from Fusarium Moniliforme Grown in Whey." *Applied and Environmental Microbiology* 41 (4): 956–58. https://www.ncbi.nlm.nih.gov/pubmed/16345758.

Mahoney, R.R. 1985. "Modification of Lactose and Lactose-Containing Dairy Products with β-Galactosidase." In *Developments in Dairy Chemistry—3*, edited by P F Fox, 69–109. Dordrecht: Springer Netherlands. doi:10.1007/978-94-009-4950-8_3.

Mahoney, Raymond R. 1998. "Galactosyl-Oligosaccharide Formation during Lactose Hydrolysis: A Review." *Food Chemistry* 63 (2): 147–54. doi:10.1016/S0308-8146(98)00020-X.

Maischberger, Thomas, Elisabeth Leitner, Sunee Nitisinprasert, Onladda Juajun, Montarop Yamabhai, Thu-Ha Nguyen, and Dietmar Haltrich. 2010. "β-Galactosidase from *Lactobacillus Pentosus*: Purification, Characterization and Formation of Galacto-Oligosaccharides." *Biotechnology Journal* 5 (8): 838–47. doi:10.1002/biot.201000126.

Mammarella, Enrique J., and Amelia C. Rubiolo. 2005. "Study of the Deactivation of β-Galactosidase Entrapped in Alginate-Carrageenan Gels." *Journal of Molecular Catalysis B: Enzymatic* 34 (1–6): 7–13. doi:10.1016/J.MOLCATB.2005.04.007.

Manera, A.P., J.C. Ores, V.A. Ribeiro, C.A.V. Burkert, and S.J. Kalil. 2008. "Optimization of the Culture Medium for the Production of β-Galactosidase from *Kluyveromyces Marxianus* CCT 7082." *Food*

Technology Biotechnology 46 (1): 66–72. http://www.ftb.com.hr/images/pdfarticles/2008/January-March/46-66.pdf.

Manera, Ana Paula, Joana Da Costa Ores, Vanessa Amaral Ribeiro, Maria Isabel Rodrigues, Susana Juliano Kalil, and Francisco Maugeri Filho. 2011. "Utilização de Resíduos Agroindustriais Em Processo Biotecnológico Para Produção de Beta-Galactosidase de *Kluyveromyces Marxianus* CCT 7082." *Acta Scientiarum. Technology* 33 (2). doi:10.4025/actascitechnol.v33i2.9919. [Use of Agroindustrial Residues in Biotechnological Process for the Production of Beta-Galactosidase from *Kluyveromyces Marxianus* CCT 7082. *Acta Scientiarum. Technology*]

Marconi, W, and F Morisi. 1979. "Industrial Applications of Fiber-Entrapped Enzymes." *Applied Biochemistry and Bioengineering* 20 (219–258): 219–58. doi:10.1016/B978-0-12-041102-3.50014-6.

Mateo, C., R. Monti, B.C.C. Pessela, M. Fuentes, R. Torres, J.M. Guisan, and R. Fernandez-Lafuente. 2004. "Immobilization of Lactase from Kluyveromyces Lactis Greatly Reduces the Inhibition Promoted by Glucose. Full Hydrolysis of Lactose in Milk." *Biotechnology Progress* 20 (4): 1259–62. doi:10.1021/bp049957m.

Matsumoto, Keisuke, Youichi Kobayashi, Natsuko Tamura, Tsunekazu Watanabe, and Tatsuhiko Kan. 1989. "Production of Galactooligosaccharides with β-Galactosidase." *Journal of the Japanese Society of Starch Science* 36 (2): 123–30. doi:10.5458/jag 1972.36.123.

Mattar, Rejane, Daniel Ferraz de Campos Mazo, and Flair José Carrilho. 2012. "Lactose Intolerance: Diagnosis, Genetic, and Clinical Factors." *Clinical and Experimental Gastroenterology* 5: 113. doi:10.2147/CEG.S32368.

Medeiros, Fabiana Oliveira de, Cristiane Reinaldo Alves, Fernanda Germano Lisboa, Daniela de Souza Martins, Carlos André Veiga Burkert, and Susana Juliano Kalil. 2008. "Ultrasonic Waves and Glass Pearls: A New Method of Extraction of Beta-Galactosodade for Use in Laboratory." *Química Nova* 31 (2): 336–39. doi:10.1590/S0100-40422008000200028.

Middelberg, Anton P.J. 1995. "Process-Scale Disruption of Microorganisms." *Biotechnology Advances* 13 (3): 491–551. doi:10.1016/0734-9750(95)02007-P.

Misselwitz, Benjamin, Daniel Pohl, Heiko Frühauf, Michael Fried, Stephan R Vavricka, and Mark Fox. 2013. "Lactose Malabsorption and Intolerance: Pathogenesis, Diagnosis and Treatment." *United European Gastroenterology Journal* 1 (3): 151–59. doi:10.1177/2050640613484463.

Mlichova, Zuzana, and Michal Rosenberg. 2006. "Current Trends of β-Galactosidase Application in Food Technology." *Journal of Food and Nutrition Research* 45: 47–54. http://www.vup.sk/en/index.php?mainID=2&navID=34&version=2&volume=45&article=783.

Najafpour, Ghasem D., Mohsen Jahanshahi, and Ghasem Najafpour. 2007. "Advanced Downstream Processing in Biotechnology." *Biochemical Engineering and Biotechnology*, 390–415. doi:10.1016/B978-044452845-2/50017-3.

Nath, Arijit, Ranjana Chowdhury, Chiranjib Bhattacharjee, and Madhumita Maitra. 2016. "Production of β-Galactosidase in a Batch Bioreactor Using Whey through Microbial Route – Characterization of Isolate and Reactor Model." *Periodica Polytechnica Chemical Engineering* 60 (4): 298–312. doi:10.3311/PPch.8286.

Nath, Arijit, Subhoshmita Mondal, Sudip Chakraborty, Chiranjib Bhattacharjee, and Ranjana Chowdhury. 2014. "Production, Purification, Characterization, Immobilization, and Application of β-Galactosidase: A Review." *Asia-Pacific Journal of Chemical Engineering* 9 (3): 330–48. doi:10.1002/apj.1801.

Navarre, W.W., and O. Schneewind. 1999. "Surface Proteins of Gram-Positive Bacteria and Mechanisms of Their Targeting to the Cell Wall Envelope." *Microbiology and Molecular Biology Reviews : MMBR* 63 (1): 174–229. http://www.ncbi.nlm.nih.gov/pubmed/10066836.

Nelson, David L., Michael M. Cox, and Albert L. Lehninger. 2013. *Lehninger Principles of Biochemistry*. New York: W.H. Freeman.

Nguyen, Thu-ha, Barbara Splechtna, Marlene Steinböck, Wolfgang Kneifel, Hans Peter Lettner, Klaus D. Kulbe, and Dietmar Haltrich.

2006. "Purification and Characterization of Two Novel β-Galactosidases from *Lactobacillus Reuteri.*" *Journal of Agricultural and Food Chemistry* 54 (14): 4989–98. doi:10.1021/jf053126u.

Nguyen, Tien-Thanh, Hoang Anh Nguyen, Sheryl Lozel Arreola, Georg Mlynek, Kristina Djinovic-Carugo, Geir Mathiesen, Thu-Ha Nguyen, and Dietmar Haltrich. 2012. "Homodimeric Beta-Galactosidase from *Lactobacillus Delbrueckii* Subsp. Bulgaricus DSM 20081: Expression in Lactobacillus Plantarum and Biochemical Characterization." *Journal of Agricultural and Food Chemistry* 60 (7): 1713–21. doi:10.1021/jf203909e.

Nivetha, A., and V. Mohanasrinivasan. 2017. "Mini Review on Role of β-Galactosidase in Lactose Intolerance." *IOP Conference Series: Materials Science and Engineering* 263: 022046. doi:10.1088/1757-899X/263/2/022046.

Nobrega, R., C.P. Borges, and A.C. Habert. 2005. "Membrane Separation Processes." In *Purification of Biotechnological Products*, 1st ed., 38–88. São Paulo: Manole.

Oehme, Felix, and Joerg Peters. 2010. "Mixed-Mode Chromatography in Downstream Process Development." *Biopharm International*, no. 3. http://www.processdevelopmentforum.com/articles/salt-tolerant-adsorption-and-unique-selectivity-are-the-major-advantages-of-mixed-mode-materials-over-single-mode-resins/.

Oliveira de Medeiros, F., C.A.B. Veiga, and S.J.Kalil. 2012. "Purification of β-Galactosidase by Ion Exchange Chromatography: Elution Optimization Using an Experimental Design." *Chemical Engineering & Technology* 35 (5): 911–18. doi:10.1002/ceat.201100571.

Ornelas, A.P., W.B. Silveira, F.C. Sampaio, and F.M.L. Passos. 2008. "The Activity of β-Galactosidase and Lactose Metabolism in Kluyveromyces Lactis Cultured in Cheese Whey as a Function of Growth Rate." *Journal of Applied Microbiology* 104 (4): 1008–13. doi:10.1111/j.1365-2672.2007.03622.x.

Pandey, Ashok. 1992. "Recent Process Developments in Solid-State Fermentation." *Process Biochemistry* 27 (2): 109–17. doi:10.1016/0032-9592(92)80017-W.

———. 2003. "Solid-State Fermentation." *Biochemical Engineering Journal* 13 (2–3): 81–84. doi:10.1016/S1369-703X(02)00121-3.

Panesar, Parmjit S., Shweta Kumari, and Reeba Panesar. 2010. "Potential Applications of Immobilized β-Galactosidase in Food Processing Industries." *Enzyme Research* 2010: 473137. doi:10.4061/2010/473137.

Panesar, Reeba, Parmjit S. Panesar, Ram S. Singh, John F. Kennedy, and Manav B. Bera. 2007. "Production of Lactose-Hydrolyzed Milk Using Ethanol Permeabilized Yeast Cells." *Food Chemistry* 101 (2): 786–90. doi:10.1016/j.foodchem.2006.02.064.

Parashar, Archana, Yiqiong Jin, Beth Mason, Michael Chae, and David C. Bressler. 2016. "Incorporation of Whey Permeate, a Dairy Effluent, in Ethanol Fermentation to Provide a Zero Waste Solution for the Dairy Industry." *Journal of Dairy Science* 99 (3): 1859–67. doi:10.3168/jds.2015-10059.

Pessoa Junior, A. 2005. "Cell Disruption." In *Purification of Biotechnological Products*, 1st ed., 7–24. São Paulo: Manole.

Raj, A.E., and G. Karanth. 2006. "Fermentation Technology and Bioreactor Design." In *Food Biotechnology*, 2nd ed. Taylor & Francis Group.

Rajoka, Muhammad Ibrahim, Khan Samia, and Riaz Sahid. 2003. "Kinetics and Regulation Studies of the Production of β-Galactosidase from Kluyveromyces Marxianus Grown on Different Substrates." *Food Technology and Biotechnology* 41 (4): 315–20. https://hrcak.srce.hr/122400.

Rao, D.G. 2009. *Introduction To Biochemical Engineering*. 2nd ed. Tata Mcgraw Hill.

Raol, Gopal G., B.V. Raol, Vimal S. Prajapati, and Nirav H. Bhavsar. 2015. "Utilization of Agro-Industrial Waste for β-Galactosidase Production under Solid State Fermentation Using Halotolerant Aspergillus Tubingensis GR1 Isolate." *3 Biotech* 5 (4): 411–21. doi:10.1007/s13205-014-0236-7.

Rech, Rosane, and Marco Antônio Záchia Ayub. 2007. "Simplified Feeding Strategies for Fed-Batch Cultivation of Kluyveromyces

Marxianus in Cheese Whey." *Process Biochemistry* 42 (5): 873–77. doi:10.1016/J.PROCBIO.2007.01.018.

Roškar, Irena, Karmen Švigelj, Mateja Štempelj, Jasna Volfand, Borut Štabuc, Špela Malovrh, and Irena Rogelj. 2017. "Effects of a Probiotic Product Containing *Bifidobacterium Animalis* Subsp. *Animalis* IM386 and Lactobacillus Plantarum MP2026 in Lactose Intolerant Individuals: Randomized, Placebo-Controlled Clinical Trial." *Journal of Functional Foods* 35: 1–8. doi:10.1016/j.jff.2017.05.020.

Rossi, A., A. Morana, I.D. Lernia, A. Di Tombrino, and M. De Rosa. 1999. "Immobilization of Enzymes on Spongy Polyvinyl Alcohol Cryogels: The Example of Beta-Galactosidase from *Aspergillus Oryzae*." *The Italian Journal of Biochemistry* 48 (2): 91–97. http://www.ncbi.nlm.nih.gov/pubmed/10434188.

Roy, Ipsita, and Munishwar N Gupta. 2002. "Downstream Processing of Enzymes/Proteins." *Proceedings of the National Academy of Sciences India* B68 (2): 175–204. https://insa.nic.in/writereaddata/UpLoadedFiles/PINSA/Vol68B_2002_2_Art01.pdf.

———. 2003. "Lactose Hydrolysis by Lactozym™ Immobilized on Cellulose Beads in Batch and Fluidized Bed Modes." *Process Biochemistry* 39 (3): 325–32. doi:10.1016/S0032-9592(03)00086-4.

Rustom, Ismail Y.S., Mervat I. Foda, and M.H. López-Leiva. 1998. "Formation of Oligosaccharides from Whey UF-Permeate by Enzymatic Hydrolysis — Analysis of Factors." *Food Chemistry* 62 (2): 141–47. doi:10.1016/S0308-8146(97)00203-3.

Sakai, Chiemi, Mari Ishida, Hideo Ohba, Hiromitsu Yamashita, Hitomi Uchida, Masao Yoshizumi, and Takafumi Ishida. 2017. "Fish Oil Omega-3 Polyunsaturated Fatty Acids Attenuate Oxidative Stress-Induced DNA Damage in Vascular Endothelial Cells." Edited by Tohru Minamino. *Plos One* 12 (11): e0187934. doi:10.1371/journal.pone.0187934.

Salton, M.R.J. 1953. "Studies of the Bacterial Cell Wall: The Composition of the Cell Walls of Some Gram-Positive and Gram-Negative Bacteria." *Biochimica et Biophysica Acta* 10 (4): 512–23. doi:10.1016/0006-3002(53)90296-0.

Sangwan, Vikas, Sudhir K. Tomar, Babar Ali, Ram R.B. Singh, and Ashish K. Singh. 2015. "Production of β-Galactosidase from *Streptococcus Thermophilus* for Galactooligosaccharides Synthesis." *Journal of Food Science and Technology* 52 (7): 4206–15. doi:10.1007/s13197-014-1486-4.

Santos, Rosângela dos, Ana Paula Resende Simiqueli, and Gláucia Maria Pastore. 2009. "Production of Galactooligosaccharide by *Scopulariopis* Sp." *Food Science and Technology* 29 (3): 682–89. doi:10.1590/S0101-20612009000300035.

Saqib, Shaima, Attiya Akram, Sobia Ahsan Halim, and Raazia Tassaduq. 2017. "Sources of β-Galactosidase and Its Applications in Food Industry." *3 Biotech* 7 (1): 79. doi:10.1007/s13205-017-0645-5.

Segel, Irwin H. 1979. *Enzyme Kinetics : Behavior and Analysis of Rapid Equilibrium and Steady State Enzyme Systems*. New York: John Wiley & Sons Inc. Publication.

Selvarajan, E., and V. Mohanasrinivasan. 2015. "Kinetic Studies on Exploring Lactose Hydrolysis Potential of β Galactosidase Extracted from *Lactobacillus Plantarum* HF571129." *Journal of Food Science and Technology* 52 (10): 6206–17. doi:10.1007/s13197-015-1729-z.

Serio, M. Di, C. Maturo, E. De Alteriis, P. Parascandola, R. Tesser, and E. Santacesaria. 2003. "Lactose Hydrolysis by Immobilized β-Galactosidase: The Effect of the Supports and the Kinetics." *Catalysis Today* 79–80: 333–39. doi:10.1016/S0920-5861(03)00059-2.

Shen, Qiuyun, Ruijin Yang, Xiao Hua, Fayin Ye, He Wang, Wei Zhao, and Kun Wang. 2012. "Enzymatic Synthesis and Identification of Oligosaccharides Obtained by Transgalactosylation of Lactose in the Presence of Fructose Using Beta-Galactosidase from Kluyveromyces Lactis." *Food Chemistry* 135 (3): 1547–54. doi:10.1016/j.foodchem.2012.05.115.

Sheu, Dey-Chyi, Shin-Yi Li, Kow-Jen Duan, and C Chen. 1998. "Production of Galactooligosaccharides by β-Galactosidase Immobilized on Glutaraldehyde-Treated Chitosan Beads." *Biotechnology Techniques* 12: 273–76. https://link.springer.com/content/pdf/10.1023/A:1008894029377.pdf.

Shin, Huyn-Jae, Jong-Moon Park, and Ji-Won Yang. 1998. "Continuous Production of Galacto-Oligosaccharides from Lactose by Bullera Singularis β-Galactosidase Immobilized in Chitosan Beads." *Process Biochemistry* 33 (8): 787–92. doi:10.1016/S0032-9592(98)00045-4.

Shukla, Triveni Piasad, and Leopold E. Wierzbicki. 1975. "Beta-galactosidase Technology: A Solution to the Lactose Problem." *C R C Critical Reviews in Food Technology* 5 (3): 325–56. doi:10.1080/10408397509527178.

Shuler, Michael L., and Fikret Kargi. 2002. *Bioprocess Engineering Basic Concepts*. 2nd ed. Upper Saddle River: Prentice Hall.

Silveira, Silvana Terra, Luci Kelin de Menezes Quines, Carlos André Veiga Burkert, and Susana Juliano Kalil. 2008. "Separation of Phycocyanin from Spirulina Platensis Using Ion Exchange Chromatography." *Bioprocess and Biosystems Engineering* 31 (5): 477–82. doi:10.1007/s00449-007-0185-1.

Silvério, Sara C., Eugénia A. Macedo, José A. Teixeira, and Lígia R. Rodrigues. 2018. "New B-Galactosidase Producers with Potential for Prebiotic Synthesis." *Bioresource Technology* 250: 131–39. doi:10.1016/j.biortech.2017.11.045.

Singhania, Reeta Rani, Anil Kumar Patel, Carlos R. Soccol, and Ashok Pandey. 2009. "Recent Advances in Solid-State Fermentation." *Biochemical Engineering Journal* 44 (1): 13–18. doi:10.1016/j.bej.2008.10.019.

Siso, M.I.G., A. Freire, E. Ramil, E.R. Belmonte, A.R. Torres, and E. Cerdán. 1994. "Covalent Immobilization of β-Galactosidase on Corn Grits. A System for Lactose Hydrolysis without Diffusional Resistance." *Process Biochemistry* 29 (1): 7–12. doi:10.1016/0032-9592(94)80053-7.

Sousa, Fani, Duarte M.F. Prazeres, and João A. Queiroz. 2008. "Affinity Chromatography Approaches to Overcome the Challenges of Purifying Plasmid DNA." *Trends in Biotechnology* 26 (9): 518–25. doi:10.1016/j.tibtech.2008.05.005.

Strzalkowska, N., K. Jasinska, and A. Jozwik. 2018. "Physico-Chemical Properties of Lactose, Reasons for and Effects of Its Intolerance in

Humans – a Review." *Animal Science Papers and Reports* 36 (1): 21–31. http://www.ighz.edu.pl/uploaded/FSiBundleContentBlockBundle EntityTranslatableBlockTranslatableFilesElement/filePath/1031/str21-32.pdf.

Subramaniyam, R., and R. Vimala. 2012. "Solid State and Submerged Fermentation for the Production of Bioactive Substances: A Comparative Study." *International Journal of Science and Nature* 3 (3): 480–86. doi:10.1021/jf980442i.

Sun, Sufang, Xuyuan Li, Shaoli Nu, and Xin You. 1999. "Immobilization and Characterization of β-Galactosidase from the Plant Gram Chicken Bean (*Cicer Arietinum*). Evolution of Its Enzymatic Actions in the Hydrolysis of Lactose." *Journal of Agricultural and Food Chemistry* 47 (3): 819–23. doi:10.1021/jf980442i.

Szczodrak, J. 2000. "Hydrolysis of Lactose in Whey Permeate by Immobilized β-Galactosidase from Kluyveromyces Fragilis." *Journal of Molecular Catalysis B: Enzymatic* 10 (6): 631–37. doi:10.1016/S1381-1177(00)00187-9.

Takeuchi, Osamu, Katsuaki Hoshino, Taro Kawai, Hideki Sanjo, Haruhiko Takada, Tomohiko Ogawa, Kiyoshi Takeda, and Shizuo Akira. 1999. "Differential Roles of TLR2 and TLR4 in Recognition of Gram-Negative and Gram-Positive Bacterial Cell Wall Components." *Immunity* 11 (4): 443–51. doi:10.1016/S1074-7613(00)80119-3.

Tanaka, A., and T. Kawamoto. 1999. *Cell and Enzyme Immobilization*. Washington, DC: The American Society for Microbiology.

Tangtua, J. 2014. "Evaluation and Comparison of Microbial Cells Disruption Methods for Extraction of Pyruvate Decarboxylase." *International Food Research Journal* 21 (4): 1331–36. http://www.ifrj.upm.edu.my/21 (04) 2014/10 IFRJ 21 (04) 2014 Tantua 716.pdf.

Taqieddin, Ehab, and Mansoor Amiji. 2004. "Enzyme Immobilization in Novel Alginate–chitosan Core-Shell Microcapsules." *Biomaterials* 25 (10): 1937–45. doi:10.1016/J.BIOMATERIALS.2003.08.034.

Tari, Canan, Fatma Isık Ustok, and Sebnem Harsa. 2009. "Optimization of the Associative Growth of Novel Yoghurt Cultures in the Production of Biomass, β-Galactosidase and Lactic Acid Using Response Surface

Methodology." *International Dairy Journal* 19 (4): 236–43. doi:10.1016/J.IDAIRYJ.2008.10.009.

Ustok, Fatma Isık, Canan Tari, and Sebnem Harsa. 2010. "Biochemical and Thermal Properties of β-Galactosidase Enzymes Produced by Artisanal Yoghurt Cultures." *Food Chemistry* 119 (3): 1114–20. doi:10.1016/j.foodchem.2009.08.022.

Uzir, M.H., and M.M. Don. 2008. *Biochemical Engineering: A Concise Introduction*. Nibong Tebal: University Sains Malaysia. http://chemical.eng.usm.my/notes/HEKARL/notes/ekc471_notes.pdf.

Vasiljevic, T., and P. Jelen. 2002. "Lactose Hydrolysis in Milk as Affected by Neutralizers Used for the Preparation of Crude β-Galactosidase Extracts from Lactobacillus Bulgaricus 11842." *Innovative Food Science & Emerging Technologies* 3 (2): 175–84. doi:10.1016/S1466-8564(02)00016-4.

Vera, Carlos, Cecilia Guerrero, and Andrés Illanes. 2011. "Determination of the Transgalactosylation Activity of Aspergillus Oryzae β-Galactosidase: Effect of PH, Temperature, and Galactose and Glucose Concentrations." *Carbohydrate Research* 346 (6): 745–52. doi:10.1016/j.carres.2011.01.030.

Viana, Caroline dos Santos, Denise Renata Pedrinho, Luiz Rodrigo Ito Morioka, and Hélio Hiroshi Suguimoto. 2018. "Determination of Cell Permeabilization and Beta-Galactosidase Extraction from Aspergillus Oryzae CCT 0977 Grown in Cheese Whey." *International Journal of Chemical Engineering* 2018: 1–6. doi:10.1155/2018/1367434.

Vinderola, C.G., and J.A. Reinheimer. 2003. "Lactic Acid Starter and Probiotic Bacteria: A Comparative 'in Vitro' Study of Probiotic Characteristics and Biological Barrier Resistance." *Food Research International* 36 (9–10): 895–904. doi:10.1016/S0963-9969(03)00098-X.

Vo, Nguyen T.K., Michael S. Mikhaeil, Lucy E.J. Lee, Phuc H. Pham, and Niels C. Bols. 2015. "Senescence-Associated β-Galactosidase Staining in Fish Cell Lines and Primary Cultures from Several Tissues and Species, Including Rainbow Trout Coelomic Fluid and Milt." *In Vitro*

Cellular & Developmental Biology - Animal 51 (4): 361–71. doi:10.1007/s11626-014-9837-z.

Vrese, Michael de, Christiane Laue, Birte Offick, Edlyn Soeth, Frauke Repenning, Angelika Thoß, and Jürgen Schrezenmeir. 2015. "A Combination of Acid Lactase from *Aspergillus Oryzae* and Yogurt Bacteria Improves Lactose Digestion in Lactose Maldigesters Synergistically: A Randomized, Controlled, Double-Blind Cross-over Trial." *Clinical Nutrition* 34 (3): 394–99. doi:10.1016/j.clnu.2014.06.012.

Vrese, Michael de, Anna Stegelmann, Bernd Richter, Susanne Fenselau, Christiane Laue, and Jürgen Schrezenmeir. 2001. "Probiotics—compensation for Lactase Insufficiency." *The American Journal of Clinical Nutrition* 73 (2): 421s–429s. doi:10.1093/ajcn/73.2.421s.

Wallenfels, Kurt, and Om Prakash Malhotra. 1962. "Galactosidases." In *Advances in Carbohydrate Chemistry*, edited by P Boyer, H Lardy, and K Myrback, 239–98. Academic Press. doi:10.1016/S0096-5332(08)60264-7.

Woudenberg-van Oosterom, M., H.J.A. van Belle, F. van Rantwijk, and R.A. Sheldon. 1998. "Immobilised β-Galactosidases and Their Use in Galactoside Synthesis." *Journal of Molecular Catalysis A: Chemical* 134 (1–3): 267–74. doi:10.1016/S1381-1169(98)00045-4.

Xavier, Janifer Raj, Karna Venkata Ramana, and Rakesh Kumar Sharma. 2018. "β-Galactosidase: Biotechnological Applications in Food Processing." *Journal of Food Biochemistry* 42 (5): e12564. doi:10.1111/jfbc.12564.

You, Sheng-ping, Xiao-nan Wang, Wei Qi, Rong-xin Su, and Zhi-min He. 2017. "Optimisation of Culture Conditions and Development of a Novel Fed-Batch Strategy for High Production of β-Galactosidase by *Kluyveromyces Lactis*." *International Journal of Food Science & Technology* 52 (8): 1887–93. doi:10.1111/ijfs.13464.

Zagustina, N.A., and A.S. Tikhomirova. 1976. "Purification and properties of beta-galactosidase from fungus, *Curvularia inaequalis*." *Biokhimiia* 41 (6): 1061–66. http://www.ncbi.nlm.nih.gov/pubmed/1051894.

Zhou, Quinn Z.K., and Dong Xiao Chen. 2001. "Immobilization of β-Galactosidase on Graphite Surface by Glutaraldehyde." *Journal of Food Engineering* 48 (1): 69–74. doi:10.1016/S0260-8774 (00)00147-3.

In: Beta-Galactosidase
Editor: Eloy Kras

ISBN: 978-1-53615-605-8
© 2019 Nova Science Publishers, Inc.

Chapter 2

SURVEY OF β-GALACTOSIDASES PROPERTIES: APPLICATIONS TO TRANSGLYCOSYLATION PROCESS

*Cecilia Porciúncula González, Cecilia Giacomini and Gabriela Irazoqui**

Laboratorio de Bioquímica, Departamento de Biociencias,
Universidad de la República, Montevideo, Uruguay

ABSTRACT

β-galactosidases (β-D-galactohydrolase, EC 3.2.1.23) are enzymes that hydrolyzes terminal β 1-4 galactosides. They belong to the GH 1, GH 2, GH 35 and GH 42 of the GH-A superfamily of glycoside hydrolases. They are widely distributed in nature and can be find in animals, plants and several microorganisms (yeast, fungi and bacteria). In nature, they function as hydrolases. In plants, they remove terminal β 1-4 galactose

---

* Corresponding Author's E-mail: mgidrv@fq.edu.uy

from polymers containing galactose, in animals and microorganisms they catalyzes the hydrolysis of lactose in galactose and glucose.

The most studied β-galactosidases are those from microorganism, such as *Escherichia coli, Aspergillus oryzae, Bacillus circulans, Streptococcus thermophilus, Kluyveromyces lactis.* This allows a collection of β-galactosidases with different properties (as optimum pH and temperature, thermal stability, stability against pH, denaturing agents, etc.), which offer a great versatility of conditions to use these enzymes for different applications.

One of the first biotechnological application of these enzymes was for lactose hydrolysis in order to produce lactose reduced milk and dairy products for consumption of lactose intolerant people. Besides, a reduced lactose content in dairy product avoids lactose crystallization improving their organoleptic properties.

Nevertheless, under adequate conditions they are able to catalyze the synthesis of oligosaccharides and galactosides. Two methods have been used for this purpose, reverse synthesis and transglycosylation. The first one involves the inversion of the hydrolysis reaction starting from a mixture of monosaccharides and nucleophilic hydroxylated molecules. This method requires high concentration of reactants as well as the presence of co-solvents in order to reduce water activity. On the other hand transglycosylation mechanisms is kinetically controlled and catalyzes the transfer of a galactosyl moiety from a galactosyl donor (lactose, *ortho*-nitrophenyl-β-D-galactopyranosyde, lactulose) to an hydroxylated nucleophile. Aliphatic alcohols, hydroxylated aminoacids such as serine and threonine, polyols, flavonoids, and monosaccharides among others have proved to be good glycosyl acceptors.

The use of the transgalactosylation mechanism is an excellent alternative to the complex chemical synthesis as it allows the synthesis of β-galactosides and β-galactooligosaccharides in a single step preserving the anomeric center configuration. Even though glycosidases are stereospecific they are not always regiospecific and so the generation of isomers mixtures could take place, depending on the structure of the acceptor molecule as well as on the source of the β-galactosidase.

Regarding synthetic applications, they have been used for enzymatic synthesis of: galactooligosaccharides with prebiotic properties as food additives; galactooligosaccharides as building blocks for synthesis of branched oligosaccharides; galactosides with potential activity as galectin inhibitors or antitumor agent; alkyl galactosides as non-ionic surfactants. β-galactosidases have also been used for galactosylation of drugs in order to improve their hydrophilicity and bioavailability.

In this chapter we will make an update regarding β-galactosidases classification, mechanisms, characterization of their properties and applications.

1. β-GALACTOSIDASE ENZYME CLASSIFICATION

Within the large family of enzymes known as glycoside hydrolases (GH), alternatively glycosidases, the β-galactosidases (β-D-galactoside-galactohydrolases) are the enzymes that catalyzes the hydrolysis of terminal non-reducing β-D-galactose residues. According to the definition given by the International Union of Biochemistry and Molecular Biology (IUBMB) Enzyme Nomenclature of the Enzyme Commission (EC) β-galactosidases are classified as EC 3.2.1.23. In this system, the EC 3.2.1 comprises the enzymes that hydrolyze O- and S-glycosyl compounds and comprises enzymes from EC 3.2.1.1 trough EC 3.2.1.196, with some deletions and reclassifications.

However, the IUBMB Enzyme nomenclature of glycoside hydrolases is based on their substrate specificity and occasionally on their molecular mechanism; such classification does not reflect the structural features of these enzymes. A classification of glycoside hydrolases in families based on amino acid sequence similarities has been proposed a few years ago by the Carbohydrate-Active Enzymes database (CAZy- http://www.cazy.org/). Since there is a direct relationship between sequence and folding similarities, such classification (i) reflects the structural features of these enzymes better than their sole substrate specificity, (ii) helps to reveal the evolutionary relationships between these enzymes, (iii) provides a convenient tool to derive mechanistic information [1,2] (iv) illustrates the difficulty of deriving relationships between family membership and substrate specificity.

To the date, there are 156 GH families (http://www.cazy.org/) reported. Because the folding of proteins is better conserved than their sequences, some of the families can be grouped in clans: (i) when new sequences are found to be related to more than one family, (ii) when the sensitivity of sequence comparison methods is increased or (iii) when structural determinations demonstrate the resemblance between members of different families [3]. To date, 18 clans named GH-A to GH-R have been reported. β-galactosidases belong to GH-A clan; the structural characteristic shared by the glycoside hydrolases that belong to GH-A clan is to have an $(\alpha/\beta)_8$- barrel as catalytic domain. The clan GH-A is compose by 21 families: 1, 2, 5, 10, 17, 26, 30, 35, 39, 42, 50, 51, 53, 59, 72, 79, 86, 113, 128, 147 and 148, and β-galactosidases is present in GH1, GH2, GH35, GH42, GH59 and GH147 families. Most studied β-galactosidases are those from families GH2 and GH35, those from eukaryotic organisms are grouped into GH35 with the exception of *Kluyveromyces lactis*, *Kluyveromyces marxianus* and *Aspergillus nidulans* β-galactosidases which belong to the GH2 family, together with prokaryotic enzymes, such as β-galactosidase from *Escherichia coli*, *Bifidobacterium bifidum*, *Lactobacillus* and *Thermotoga maritima*.

2. SOURCES AND FUNCTIONS OF β-GALACTOSIDASE

β-Galactosidase enzyme is present in a wide multiplicity of organisms including plants, animal tissues and microorganisms. As it was already mentioned β-galactosidases catalyze the hydrolysis of terminal non-reducing β-D-galactose residues. The best-known biological function of β-galactosidases is to hydrolyze lactose, and hence its common way of calling it lactase. However, as can be seen, the definition is much wider than the hydrolysis of the specific substrate lactose. While all lactases are β-galactosidases, the converse is not true since most β-galactosidases from plant cells and mammalian organs (other than the intestine) have little or no activity on lactose, which is not surprising since this substrate cannot be encountered in their native environment.

β-Galactosidase action has been associated with the release of energy for rapid growth (lactose hydrolysis in mammals and bacteria, xyloglucan mobilization in cotyledons), release of free galactose during normal metabolic recycling of galactolipids, glycoproteins, and cell wall components, and degradation of cell wall components during senescence [4–11].

2.1. Animal β-Galactosidases

In the case of animal tissues, two classes of β-galactosidase have received special attention because of their physiological importance: the enzymes acting on complex galactosylsphingolipids found in brain and intestine, and the intestinal lactases because of their involvement in hydrolysis of dietary disaccharides lactose.

Lactase responsible for cleaving lactose into its constituent absorbable monosaccharides, glucose and galactose, is essential for the nourishment of new-born mammals, whose sole source of nutrition is milk, in which lactose is the major carbohydrate component [12]. In most mammals, including humans, lactase expression decreases after the weaning period is over, when it is no longer needed, since lactose is not found elsewhere in the diet. Undigested lactose passing through the small intestine into the colon has two physiological effects: first, an osmotic gradient is set up across the gut wall, which results in an influx of water, causing symptoms of diarrhoea. Second, the lactose can be fermented by colonic bacteria, to produce fatty acids and gaseous by-products (including hydrogen, used in the tolerance test), potentially causing discomfort, bloating and flatulence. However most lactase non-persistent individuals can tolerate small amounts of lactose (as in tea or coffee), and some can consume a lot without ill effects [13, 14]. Variation in the composition of the gut flora between individuals [15], as well as a psychosomatic component [16] may account for some of the interindividual variation in symptoms [17].

Table 1. β-galactosidases from animal described by the CAZy database

Organism scientific name	Common name
Canis lupus familiaris	dog
Felis catus	domestic cat
Homo sapiens	human
Mus musculus	house mouse

In relation to the others β-galactosidases that are not lactases, there is profuse literature because deficiencies in their enzymatic activities is closely related to the manifestation of some diseases. Galactosylceramidase, which catalyzes the hydrolysis of galactosylceramide to ceramide and galactose, also acts on galactosylsphingosine (psychosine), lactosylceramide, and monogalactosyldiglyceride [18]. The activity of this enzyme is deficient in globoid cell leukodystrophy (GLD) occurring in humans, dogs, and mice. Another example is the G_{M1} ganglioside β-galactosidase which is active toward G_{M1} ganglioside, asialo-G_{M1} ganglioside, lactosylceramide, and glycoproteins with terminal galactose, which deficiency causes G_{M1} gangliosidosis [18].

Table 1 shows β-galactosidases from animal tissues described by the CAZy database, all of them belong to GH Family 35.

2.2. Plants β-Galactosidases

In the plant kingdom, β-galactosidases can be found in several plant tissues and are known to attack different types of substrates such as glycoproteins, glycolipids, flavonoids and alkaloids. β-galactosidase was first detected in plants in emulsions of Rosaceae (peaches, apricot and apple) by Bourquelot and Herissey in 1903 (cited by Wallenfels et al. (1961) [19]). The biological roles of these enzymes include degradation of structural polysaccharides in plant cell walls, thereby controlling fruit softening during ripening [20, 21].

The role of β-galactosidases in tomato fruit has been deeply studied and has resulted from physiological and biochemical data showing that galactose is the most dynamic sugar residue of the cell wall during tomato fruit development. In particular, these physiological studies showed that there was a significant net loss of galactosyl residues from the wall throughout fruit development, and the rate of galactosyl residue release increased during ripening. It was also shown that free galactose levels, although stable throughout the preripening stages of fruit development, increased rapidly during ripening [22, 23].

Table 2 shows β-galactosidases from plants described by the CAZy database, all of them belong to GH Family 35. β-galactosidase is widely distributed in different plant tissues, such as leaves [24], seedlings [25], hypocotyls [26], and meristem zones of roots, cotyledons, vascular tissues, trichrome, and pollens [27].

Table 2. β-galactosidases from plants described by the CAZy database

Organism scientific name	Organism common name	Organism scientific name	Organism common name
Arabidopsis thaliana	Mouse-ear cress	Malus domestica	Apple
Asparagus officinalis	Garden asparagus	Mangifera indica	Mango
Brassica oleracea	Cabbage, broccoli, cauliflower, kale, Brussels sprouts, collard greens, savoy, kohlrabi, and gai lan.	Oryza sativa Japonica Group	Asian rice short-grained japonica variety
Capsicum annuum	Bell peppers, jalapeños, New Mexico chile, and cayenne peppers	Persea Americana	Avocado
Carica papaya	Papaw or pawpaw	Pyrus pyrifolia	Pear
Cicer arietinum	Chickpea	Raphanus sativus	Radish
Citrus sinensis	Sweet orange	Sandersonia aurantiaca	Christmas bell
Coffea arabica	Arabian coffee	Solanum lycopersicum	Tomato
Dianthus caryophyllus	Carnation	Tropaeolum majus	Garden nasturtium, Indian cress, or monks cress
Fragaria x ananassa	Strawberry	Vigna radiata	Mung bean
Linum usitatissimum	Flax or linseed	Vitis vinifera	Grape vine
Lupinus angustifolius	Narrow-leaved lupine and blue lupine		

Most plant β-galactosidases contain C-terminal domains that are homologous to animal lectins and have thus been described as lectin-like domains. For example, fifteen rice (Oryza sativa L.) β-galactosidase (OsBGal) genes were identified, and twelve of these encode proteins that contain the galactose-lectin-like domain, including OsBGal1 [28].

2.3. Microorganisms β-Galactosidases

β-Galactosidase occurs in a variety of microorganisms, including yeasts, fungi and bacteria [29]. In microorganisms, the enzyme has lactase activity, since they use lactose as the carbon source β-galactosidases from fungi and yeast are part of GH1, GH2 and GH35 families, and among them the most outstanding ones are the *Kluyveromyces, Aspergillus, Penicillium* and *Trichoderma genus* (Table 3).

Table 3. β-Galactosidases from fungi and yeast described by the CAZy database

GH Family	Yeast
GH2	*Kluyveromyces lactis*
GH2	*Kluyveromyces marxianus*
GH Family	**Fungi**
GH1	*Hamamotoa singularis*
GH35	*Aspergillus candidus*
GH35	*Aspergillus niger*
GH35	*Aspergillus oryzae*
GH35	*Aspergillus phoenicis*
GH35	*Bispora*
GH35	*Penicillium canescens*
GH35	*Penicillium expansum*
GH35	*Talaromyces aerugineus*
GH35	*Trichoderma reesei*

Table 4. β-galactosidases from bacteria described by the CAZy database

GH Family	Bacteria name	GH Family	Bacteria name
GH2	Actinobacillus pleuropneumoniae	GH2	Lactobacillus delbrueckii
GH35	Akkermansia muciniphila	GH2	Lactobacillus lactis
GH42	Alicyclobacillus acidocaldarius subsp. Acidocaldarius	GH2	Lactobacillus plantarum
GH2	Arthrobacter psychrolactophilus	GH2	Lactobacillus reuteri
GH35/GH42	Arthrobacter sp.	GH2	Lactobacillus sakei
GH2/GH35/GH42	Bacillus circulans	GH2	Mannheimia succiniciproducens
GH42	Bacillus coagulans	GH42	Paenibacillus sp.
GH42	Bacillus licheniformis	GH35	Paenibacillus thiaminolyticus
GH2	Bacillus megaterium	GH2	Pantoea agglomerans
GH42	Bacillus subtilis	GH2	Paracoccus sp
GH2/ GH147	Bacteroides ovatus	GH42	Planococcus sp.
GH2/GH35	Bacteroides thetaiotaomicron	GH42	Pseudarthrobacter chlorophenolicus
GH42	Bifidobacterium adolescentis	GH2	Pseudoalteromonas haloplanktis
GH42	Bifidobacterium animalis	GH2	Psychromonas marina
GH2/GH42	Bifidobacterium bifidum	GH42	Rahnella sp.
GH2/GH42	Bifidobacterium breve	GH1/GH59	Ruminiclostridium cellulolyticum
GH2/GH42	Bifidobacterium longum	GH2	Saccharopolyspora rectivirgula
GH2	Caldicellulosiruptor lactoaceticus	GH2	Serratia sp.
GH42	Caldicellulosiruptor sp	GH2	Sinorhizobium meliloti
GH35/ GH42	Carnobacterium maltaromaticum	GH2	Staphylococcus xylosus
GH35	Cellvibrio japonicus	GH2/GH35	Streptococcus pneumoniae
GH2	Clostridium acetobutylicum	GH35	Streptococcus suis

Table 4. (Continued)

GH Family	Bacteria name	GH Family	Bacteria name
GH42	Clostridium cellulovorans	GH2	Streptococcus thermophilus
GH42	Clostridium perfringens	GH2	Thermoanaerobacter mathranii
GH42	Deinococcus geothermalis	GH2	Thermoanaerobacter pseudethanolicus
GH42	Dickeya dadantii	GH2	Thermoanaerobacterium thermosulfurigenes
GH42	Dictyoglomus turgidum	GH2/GH42	Thermotoga maritima
GH2	Enterobacter cloacae	GH42	Thermotoga naphthophila
GH2	Escherichia coli	GH2/GH42	Thermotoga neapolitana
GH2	Flavobacterium sp.	GH42	Thermus oshimai
GH1	Geobacillus kaustophilus	GH1/GH2/GH42	Thermus thermophilus
GH42	Geobacillus stearothermophilus	GH1	uncultured Meiothermus sp.
GH42	Geobacillus thermocatenulatus	GH2	Vibrio sp.
GH42	Isoptericola sp.	GH2	Vibrio vulnificus
GH2/GH42	Klebsiella pneumoniae	GH2/GH35	Xanthomonas campestris pv. Campestris
GH2	Lactobacillus fermentum	GH35	Xanthomonas phaseoli pv. Manihotis
GH2/GH42	Lactobacillus acidophilus	GH2	Yersinia pestis

For the case of β-galactosidases from bacteria, they are present in the six families of glycosyl hydrolases that include β-galactosidases. To date, β-galactosidases belonging to at least 45 genus of bacteria have been reported in the CAZy database (Table 4).

Tables 1 to 4 show the diversity of origins of enzymes with β-galactosidase activity. However, the easier manipulation of microorganisms and the better yields obtained in the purification of their enzymes have favored their establishment as a main source for industrial production of β-galactosidases. Although bacteria could offer more versatility, the corroborated GRAS (Generally Recognized As Safe) status of yeasts like *K. lactis* and *K. marxianus*, and of fungi like *A. niger* and *A. oryzae*, still places them among the favorite sources of β-galactosidase for food biotechnology and pharmaceutical industry [30].

3. β-GALACTOSIDASES STRUCTURES

3.1. *Escherichia Coli* β-Galactosidase Structure

E. coli β-galactosidase is one of the most studied and best-characterized enzymes in its class. Due to its wide and rich history, both in the field of molecular biology and biochemistry, it has become a valuable tool in these areas. Already in 1964, Karlsson et al. [31] reported that it was a tetrameric enzyme, which was confirmed in 1992 by Jacobson and Matthews through X-ray studies [32]. The purification of the enzyme was first carried out in the early 1950s by Cohn and Monod. In the 1970s, Fowler and Zabin determined its amino acid sequence by chemical sequencing [33]. The enzyme is the product of the Z gene which belongs to the lac operon, and so it is called the lacZ gene [34].

This tetrameric enzyme is composed of 4 polypeptide chains, A-D, of 1023 amino acids per chain. Each monomer is composed of 5 domains, 1-5 (Figure 1). Domain 3 has a structure $(\alpha/\beta)_8$ or TIM barrel, within which the active site of the enzyme is located, in the loops that connect the terminal carboxyl of the β strands with the alpha helices.

In the tetramer, the four monomers are grouped around three axes of symmetry perpendicular to each other. It can be considered that each of these axes forms a different interface between two pairs of different monomers. The "long" interface, relates the monomers A with B, and C with D. The "active" interface relates A to D, and B to C.

Finally, a much smaller interface, relates A with C and B with D [35]. The active site is constituted from loops that come from the first and fifth domain of the same monomer and by a loop that comes from domain 2 of another monomer, and form a deep groove that penetrates into the heart of TIM barrel domain 3. Monomer A donates the loop of its domain 2 to complete the active site of monomer D, and vice versa, a loop of monomer D completes the active site of monomer A. Reciprocally there is a donation and acceptance between monomers B and C in order to obtain four functional active sites. Each active site is well separated from the other three and probably acts independently, but since two monomers are needed to complete an active site, the individual monomers would not be active.

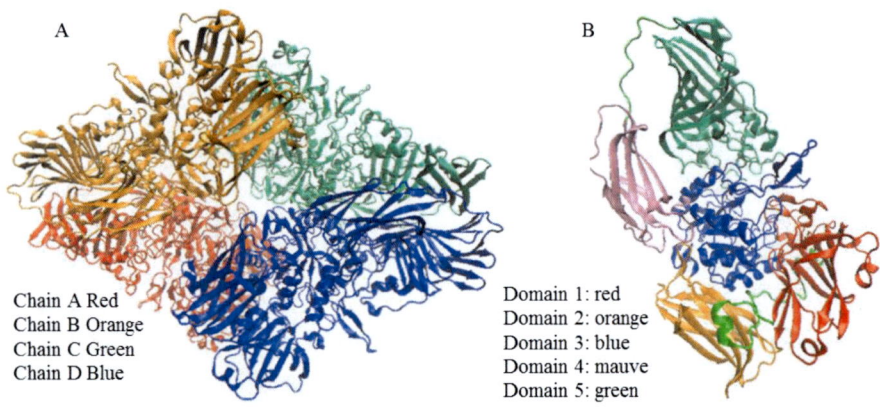

Figure 1. *Escherichia coli* β-galactosidase structure. A) View of the tetramer (PDB ID: 1JYN); B) view of the A chain with the 5 domains indicated in different colors (PDB ID: 1JYN, Chain A) [36].

β-Galactosidase requires Mg^{+2} or Mn^{+2} for full catalytic activity, but the exact role of this ion in catalysis is unclear. The active site also includes a monovalent cation (usually either Na^+ or K^+) important for the enzymatic activity. The two ion sites are situated a few A° apart from the active site, both very near to the interface between two domains of the protein [37].

3.2. *Aspergillus Oryzae* β-Galactosidase Structure

In the case of the *A. oryzae* β-galactosidase, the amino acid sequence includes 1005 residues with an average molecular mass of 110 kDa [38]. Maksimainen et al. reported in 2013 [39] the crystal structure of *A. oryzae* β-galactosidase that consists of six domains. The first domain (Asp40-Thr397) is the catalytic domain with $(\alpha/\beta)_8$ barrel fold, then it has a second (Ala398-Val475), third (Asp476-Tyr571) and fourth domain (Asn572-Ala664). The fifth domain is built from two pieces of the polypeptide chain (Pro665-Leu683) and (Tyr859-Tyr1005), and the sixth domain (Pro684-Leu858) is inserted between the pieces of the fifth domain. Domains 2-6 form a horseshoe with five anti-parallel β-sandwich structures around the catalytic domain (Figure 2).

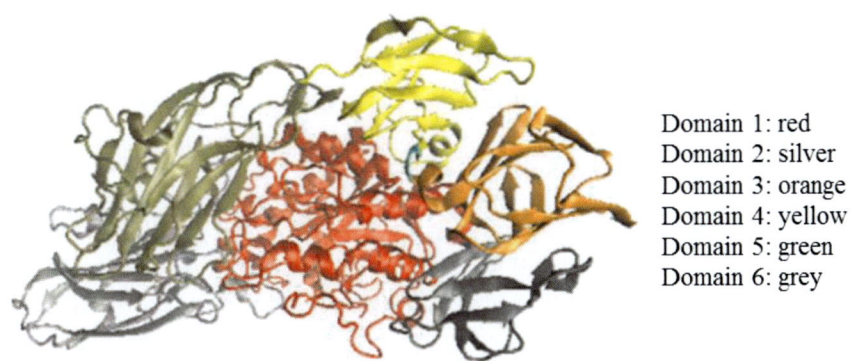

Domain 1: red
Domain 2: silver
Domain 3: orange
Domain 4: yellow
Domain 5: green
Domain 6: grey

Figure 2. *Aspergillus oryzae* β-galactosidase structure. View of the monomer with the 6 domains indicated in different colors (PDB ID: 4IUG) [36].

According to its sequence, *A. oryzae* β-galactosidase contains 11 putative N-glycosylation sites. Based on the crystal structure, only one of them, Asn805 is located inside the protein and all the remaining Asn residues, which constitutes putative glycosylation sites are located on the surface of the protein. Glycans were observed in six places. However, the observed glycan structures do not necessary resemble the native *A. oryzae* β-galactosidase glycans because *A. oryzae* β-galactosidase was heterologously expressed in the yeast *K. lactis*. Three carbohydrate chains have large so-called high-mannose structures that are especially located in the interfaces of domains. The locations of the large oligosaccharide chains indicate that N-glycosylation stabilizes the structure of *A. oryzae* β-galactosidase with several hydrogen bond interactions. Another role for the glycans can also be a protection from proteolysis. For example, pepsin cleaves peptide bonds between aromatic and hydrophobic residues. There are several hydrophobic and aromatic residues (Trp307, Tyr316, Ala317, Ala770, Ala773, Tyr772, and Val937) under the high-mannose chains beginning from Asn622 and Asn914.

3.3. *Kluyveromyces Lactis* β-Galactosidase Structure

K. lactis β-galactosidase forms a homo-oligomer of four subunits that can be described as a dimer of dimers. Each chain (A–B–C–D) consists of 1024 residues with a molecular mass of 119 kDa as calculated from its primary structure. Each subunit folds into five domains, only one with assigned catalytic function. Domain 1 (residues 32–204) presents a jelly roll fold classified as sugar binding domain. Domains 2 (residues 205–332) and 4 (residues 643–720) form two immunoglobulin-like β-sandwich domains. Domain 3 (residues 333–642) folds into a TIM barrel domain harboring the catalytic pocket and domain 5 (residues 741–1025) is small chain. There are two extended regions of the protein that cannot be assigned to any of the domains. One is the N-terminal region (residues 2–31) and the other is a small solvent exposed linker that connects domains 4 and 5 (residues 721–740) (Figure 3) [40].

The *K. lactis* β-galactosidase was found to be tetrameric in the crystal, with the four molecules building up the asymmetric unit. Several studies have reported the presence of two active forms in native electrophoresis analysis of β-galactosidase samples purified from *K. lactis*, which were attributed to the presence of dimers and tetramers [41]. The fact that the oligomerization pattern observed in the crystal corresponds to a "dimerization of dimers" is consistent with the experimental results [40]. The catalytic residues were identified as Glu482 and Glu551, located in a pocket found at one side of the TIM barrel domain, in the center of each monomer. The catalytic pocket is surrounded by residues from domains 1, 3 and 5, that shape a very narrow cavity about 20 Å deep. One magnesium and two sodium ions are located at the active site in the galactose complex structure. The magnesium and one of the sodium ions were found close to the galactose ring of the substrate, resembling the metal binding scheme of *E. coli* β-galactosidase catalytic site [37]. A second sodium ion was found filling a gap left by the shorter side chain of residue Trp190.

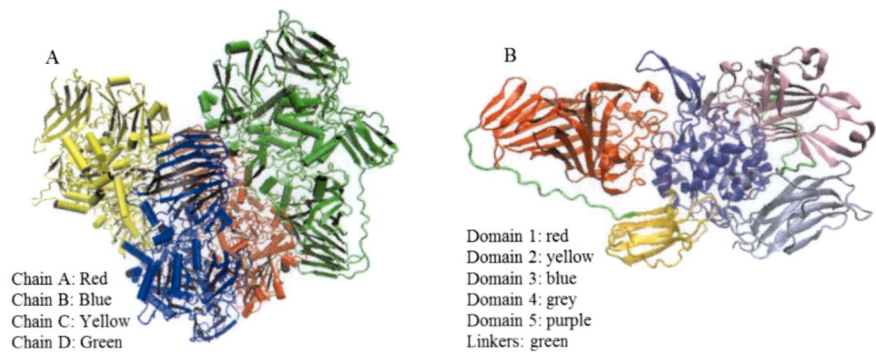

Figure 3. *Kluyveromyces lactis* β-galactosidase structure. A) View of the dimer of dimers (PDB ID: 3OB8; B) view of the A chain with the 5 domains indicated in different colors (PDBID: 3OB8, Chain A) [36].

4. MICROBIAL β-GALACTOSIDASES PROPERTIES

Properties, specificity and structure of β-galactosidase significantly differ on the microbial source of the enzyme, allowing a collection of β-galactosidases with different characteristics, which offer a great versatility of conditions to use these enzymes for different applications. Table 5 shows the properties of a variety of microbial β-galactosidases. The enzymes from yeasts have a range of pH optimum of 6.5–7.0 and are called neutral lactases. On the other hand, the fungal β-galactosidases with a range of optimum pH of 3–5 are more suitable for acidic reaction medium [42, 43]. The activity of different β-galactosidases also depends on presence of ions. The fungal β-galactosidases are active without ions as cofactors, the yeast β-galactosidase isolated from *K. lactis* requires ions, such as Mn^{2+}, Na^+, or K^+ [44] and β-galactosidase from *K. fragilis* ions such as Mn^{2+}, Mg^{2+}, K^+ [45]. On the contrary, Ca^{2+} and heavy metals inhibit the enzyme activity of all β-galactosidases [46, 47].

Many β-galactosidases have product inhibition, those of fungi are more sensitive to galactose, and those of yeasts by both, galactose and glucose. In the *A. oryzae* enzyme the product of lactose hydrolysis galactose inhibited competitively the enzyme activity, whereas glucose did not [48]. For *K. lactis* β-galactosidase galactose and glucose are competitive and no competitive inhibitor respectively [44].

Table 5. Microbial β-galactosidases properties

	Microorganism	T optimum [°C]	pH optimum	References
Yeasts (intracellular)	*Kluyveromyces lactis*	30–35	6,5–7,0	[49–51]
	Kluyveromyces fragilis	30–35	6,6	[29, 52]
Bacteria (intracellular)	*Escherichia coli*	40	7,2	[19, 53, 54]
	Lactobacillus thermophilus	55	6,2	[29, 55]
	Leuconostoc citrovorum	66	6,5	[56]
	Bacillus circulans	65	6,0	[52]
Fungi (extracellular)	*Aspergillus niger*	55–60	3,0–4,0	[57]
	Aspergillus oryzae	50–55	5,0	[48, 58]

5. CATALYTIC MECHANISM OF β-GALACTOSIDASES

As previously mentioned, galactosidases are classified as hydrolases (EC 3.2.1.23) and catalyze the hydrolysis of β-D-galactosides, particularly lactose (galactopyranosyl β-D-(1→4)-glucopyranoside), fulfilling the following general equation.

$$A + B \rightleftharpoons P + Q \tag{1}$$

Where A can be lactose, B water and P and Q glucose and galactose respectively. Equation (1) represents enzymatic reactions with more than one substrate. Nevertheless, as the second substrate is water, always present in large excess in aqueous medium, its concentration can be considered constant, and consequently galactosidases are treated as single substrate enzymes. In spite of this, in the presence of a nucleophile different from water they can function as transferases following a substituted enzyme mechanism (also known as double-displacement or Koshland retaining mechanism) (Figure 4) [59, 60].

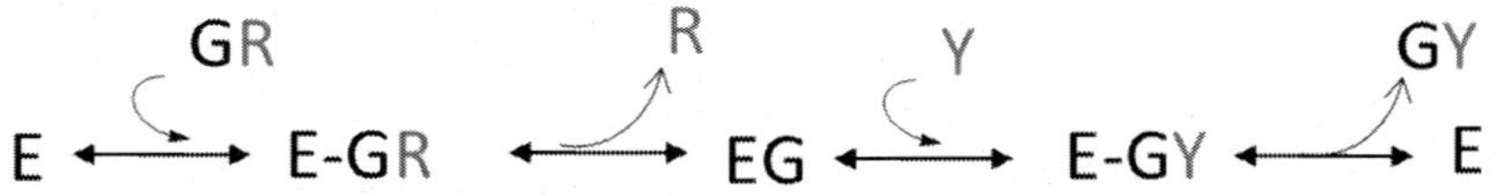

$$E \xrightleftharpoons{GR} E\text{-}GR \xrightleftharpoons{R\uparrow} EG \xrightleftharpoons{Y} E\text{-}GY \xrightleftharpoons{GY\uparrow} E$$

Figure 4. Substituted enzyme mechanism catalyzed by β-galactosidase. E: Enzyme, G: Galactose, R: aglicone of the galactosyl donor, Y: molecule acceptor.

This mechanism takes place in two steps. First, a galactose donor reacts with the enzyme forming a galactosyl-enzyme intermediate (EG) with the concomitant release of the aglicone of the galactose donor (R). In a second step a nucleophile (Y) reacts with the galactosyl-enzyme complex forming a new galactoside [61, 62]. In nature, when lactose is the galactose donor and water the nucleophile, lactose hydrolysis takes place and glucose is released in the first step, while galactose is released in the second step. Like most glycosidases, β-galactosidases preserve the stereochemistry of the anomeric center allowing the synthesis of anomerically pure galactosides in a single step, avoiding several protection and deprotection steps necessary in chemical synthesis [61, 63, 64, 65].

β-galactosidase catalytic mechanism is similar to that of other retaining glycosidases and is illustrated in Figure 5.

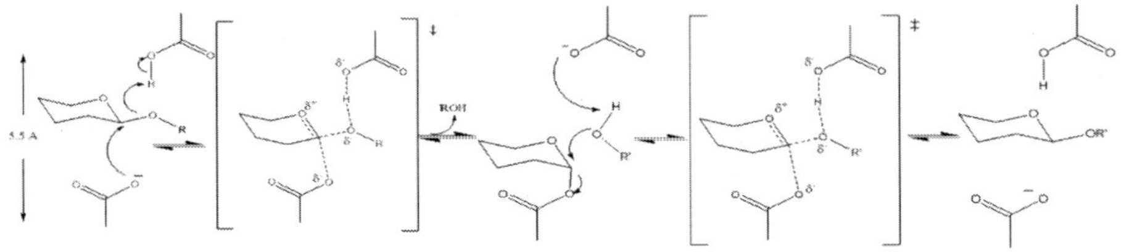

Figure 5. Catalytic mechanism of β-galactosidase.

The active site involves two glutamic acid residues. One of them is protonated and assists the release of the aglicone of the galactosyl donor by the transference of a proton to the oxygen of the anomeric carbon, leading to the formation of an enzyme-oxocarbenium ion intermediate. Simultaneously the second carboxylate, which is deprotonated in the enzyme native state, stabilizes the oxonium ion. Then, either a water molecule or an alternative nucleophile carries on a nucleophilic attack, with the consequent formation of new glycoside. The nucleophilic attack takes place in the same face of the glycosyl-enzyme intermediate from which the leaving group was released, resulting in a retention of the configuration of the anomeric center [61-63, 66, 67].

6. TRANSGALACTOSYLATION SYSTEM

The generation of galactosides catalyzed by β-galactosidases when they function as transferases is known as transglycosylation process to differentiate it from the reverse synthesis. The latter consists in the reversion of the thermodynamic equilibrium either by increasing the hydrolysis product concentration (i.e., galactose and glucose or the corresponding aglycone), or by reducing water activity using organic solvents. This process is thermodynamically controlled and is not very successful due to the low solubility of monosaccharides in organic solvent [60-63, 68]. Besides, it is necessary either to stabilize the enzyme, or use a stable β-galactosidase, so that it can tolerate the organic solvent.

On the other hand, transgalactosylation process is kinetically controlled and involves several secondary reactions that compete with the transgalactosylation reaction (Figure 6).

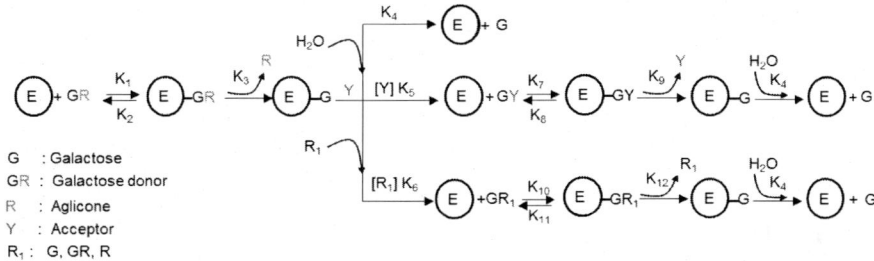

G : Galactose
GR : Galactose donor
R : Aglicone
Y : Acceptor
R_1 : G, GR, R

Figure 6. Transgalactosylation scheme.

The galactose donor (GR), as well as the aglycone (R), if it reaches an adequate concentration, can also works as acceptor generating a new galactoside or galactooligosaccharide (GR_1). Besides, the synthesized galactosides (GY or GR_1) can also be enzyme substrate and so prone to enzyme hydrolysis. So, the yield of the main galactoside (GY) is determined by the balance between the rates of all the reactions involved in the transgalactosylation mechanism. In this context, reaction time is a critical parameter, which should be optimized. Transgalactosylation process generally allows higher reaction yields than reverse synthesis, because it is possible to overcome equilibrium concentrations. Moreover, is faster and can be achieved in a few hours, while reverse synthesis needs days. Other parameters that affect the performance of transgalactosylation systems are the concentration and nature of donor and acceptor molecules, as well as the enzyme source and will be further described [63, 69, 70].

6.1. Influence of Donor Nature and Concentration in the Performance of the Transgalactosylation Synthesis

The structure of the galactose donor is relevant for the kinetic control of the transgalactosylation reaction. Donors with low K_M binds tightly to the enzyme, reducing product hydrolysis. Moreover, fast-reacting donor keeps the reaction time short decreasing time available for product hydrolysis [63, 71]. Indeed, Irazoqui et al. reported in 2013 a reduction on the optimal reaction time for the synthesis of 2-O-β-D-galactopyranosyl-ethyleneglycol catalyzed by *A. oryzae* β-galactosidase from 24 hours to 4 hours when using *ortho*-nitrophenyl-β-D-galactopyranoside (ONPG, 50 mM) instead of lactose (70 mM). In both cases, the transglycosylation reaction was performed at 25°C, ethyleneglycol 9.0 M as acceptor, with a maximum reaction yield of 70%. This is consistent with the difference in *A. oryzae* K_M for ONPG (2 mM determined at 25°C) and lactose (54 mM determined at 50°C) [71].

The galactose donor concentration is also relevant in the performance of the transgalactosylation system. High concentrations of galactose donor increases transgalactosylation yields as water activity is lowered, reducing the hydrolysis reaction and shifting the equilibrium towards transgalactosylation. Besides, lactose competes with the transgalactosylation product for the enzyme, shortening its hydrolysis [72]. An increase in the amount of produced galactosides due to increase in lactose concentration has been reported [73, 74].

6.2. Influence of Acceptor Nature and Concentration in the Performance of the Transgalactosylation Synthesis

Several molecules have been used as acceptors in the transgalactosylation system such as alcohols, polyols, alcohol amines, catechols or sugars. Among them the most studied are aliphatic alcohols and saccharides. As monosaccharides are polyhydroxylated molecules, mixtures of regioisomers can be obtained. Transgalactosylation yields increases when the anomeric carbon of the acceptor is linked to a molecule containing a phenyl group, and regioselectivity is strongly dependent on the nature of the atom linked to the anomeric carbon [70, 75, 76]. In fact, *E. coli* β-galactosidase shows specificity for the generation of β-(1→4) and β-(1→3) galactosidic linkages when using substituted xylopyranoses as acceptor and ONPG as galactosyl donor. While this enzyme is regioselective for the synthesis of β-D-(1→4)-galactopyranosyl-D-methylxylopyranoside, a mixture of β-D-(1→3)-galactopyranosyl-D-benzylxylopyranoside and β-D-(1→4)-galactopyranosyl-D-benzylxylopyranoside (84:16) is achieved [70]. In the same context, *B. circulans* β-galactosidase predominantly forms β-(1→4) linkages and to a lesser extent β-(1→6) linkages, when lactose is used as donor and N-acetylglucosamine or N-acetylgalactosamine as acceptors. Nevertheless, β,β-(1→1) and β-(1→6) linkages are formed when the acceptor is 3-acetamido-3-deoxy-D-glucose, and a mixture of β-(1→4), β-(1→3) and β-(1→6) with ratios of (45:37:18) or (83:14:3) when methyl-β-D-glucoside or methyl-β-D-galactoside are used respectively [77].

Generation of β-(1→3) linkages prevails over β-(1→6) isomers when *B. circulans* β-galactosidase catalyze the transfer of galactose from ONPG to α-D-N-acetylgalactopyranosyl-p-nitrophenol. The products β-D-galactopyranosyl-(1→3)-α-D-N-acetylgalactopyranosyl-p-nitrophenol and β-D-galactopyranosyl-(1→6)-α-D-N-acetylgalactopyranosyl-p-nitrophenol are obtained with a yield of 79.1% and a ratio of (96:4). In the same way β-D-galactopyranosyl-(1→3)-α-D-N-acetylglucopyranosyl-p-nitrophenol is synthesized using α-D-N-acetylglucopyranosyl-p-nitrophenol as acceptor. However, the regioselective synthesis of the β-(1→6)-linked disaccharide β-D-galactopyranosyl-(1→6)-β-D-galactopyranosyl-p-nitrophenol takes place when β-D-galactopyranosyl-p-nitrophenol is used as acceptor. In the case of the α-D-galactopyranosyl-p-nitrophenol acceptor, the β-(1→6)-linked disaccharide β-D-galactopyranosyl-(1→6)-α-D-galactopyranosyl-p-nitrophenol and the β-(1→3)-linked one are formed with a molar ratio of (19:6) [78].

However, the anomeric configuration of the acceptor do not determine the regioselectivity of *B. circulans* β-galactosidase when ethylthio-monosaccharides are used as acceptors. Mainly β-(1→4) linkages are formed when ethyl-1-thio-β-D-glucopyranoside, 1-thio-β-D-galactopyranoside or ethyl-2-acetamido-2-deoxy-l-thio-β-D-glucopyranoside are used as acceptors [79].

Regarding aliphatic alcohols as acceptors, most β-galactosidases show preferences for primary alcohols as they are more reactive [73, 76, 80].

Contrary to what may be expected, not any hydroxylated molecule is a good acceptor for the transgalactosylation system. Phenols are reactive leaving groups and hence, weak acceptors inefficiently glycosylated by glycosidases. This might be related to the low nucleophilicity of phenolic hydroxyl groups [63, 81]. In fact, our research group tried unsuccessfully to galactosylate tyrosine using *A. oryzae* β-galactosidase and lactose as donor. Even though synthesis of 4-hydroxy-phenyl-β-D-galactopyranoside (hydroquinone galactoside) with β-galactosidases from *K. lactis*, *A. oryzae*, *B. cirulans* and *Thermus sp* has been reported by Kim et al. in 2010, the highest amount of galactoside was of 2.29 mM (obtained with *K. lactis*) starting from 100 mM lactose and 100 mM hydroquinone [82].

Nevertheless, when the hydroxyl group is separated from the phenol moiety by one or two carbons as in benzylic alcohol or tyrosol, enzymatic galactosylation by transglycosylation process using lactose can be achieved with *E. coli* and *A. oryzae* β-galactosidases [83, 84]. Other bad acceptors for the transgalactosylation system catalyzed by *A. oryzae* β-galactosidase are hydroxyurea and lactic acid (Figure 7). Probably the closeness of a carbonyl group to the hydroxyl group reduce the nucleophilicity and so the reactivity of the hydroxyl group [85]. In fact the esterification of the carboxylic group of the lactic acid with ethanol to form ethyl-L-lactate (Figure 7) generates a good acceptor for the transgalactoyslation system as reported by Jia et al. (2007) [86].

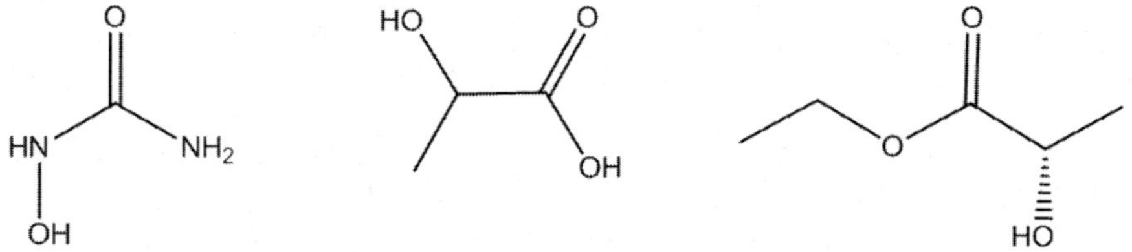

Hydroxyurea   Lactic Acid   Ethyl Lactate

Figure 7. Structure of other possible acceptor molecules.

High acceptor concentrations reduce water activity favoring transglycosylation over hydrolysis, so it has been frequently proposed as a tool for increasing synthesis yield in the enzymatic production of galactosides, provided the acceptor is not an enzyme inhibitor. This point would be further discussed in following sections [70, 71, 73, 74].

6.3. Influence of the Enzyme Source

Unlike glycosyltransferases, glycosidases have a low selectivity towards the acceptor molecules, leading to the synthesis of a mixture of isomers with poly-hydroxylated acceptors. In general β-galactosidases show preferences for primary hydroxyl groups [73]. Thus in the case of the saccharide acceptor β-(1→6) glycosidic linkages are formed due to the more reactive primary hydroxyl function [75, 84]. However, regioselectivity varies depending on the enzyme source [61, 63].

E. coli β-galactosidase shows preference for primary towards secondary hydroxyl groups when alcohols and diols were used as acceptors. The preference was for the R-enantiomer in the case of chiral alcohols and diols [76, 80]. When using sugars as acceptors, this enzyme shows preference for β -(1→6) linkages [87, 88]. Nevertheless, as mentioned in the previous section, specificity could be altered when using monosaccharides substituted in the anomeric center, such as methyl or benzyl xylose, where E. coli β-galactosidase is specific for β - (1→4) and β - (1→3) glycosidic linkages [70].

A. *oryzae* β-galactosidase also shows preference for primary alcohols as well as for the generation of β-(1→6) linkages in the case of saccharide acceptors [73, 80, 84]. Besides, when pro-chiral acceptors such as glycerol or erythritol are galactosylated, (1:1) racemic mixtures of the respective optical isomers, ((2R) and (2S)-3-O-β-D galactopyranosyl-glycerol, (2R,3S) and (2S,3R)-4-O-β-D-galactopyranosyl-erythritol), are generated respectively [89].

Unlike *E. coli* and *A. oryzae* β-galactosidases, the enzyme from *B. circulans* shows preference for secondary alcohols as well as for the formation of β-(1→4) in addition to β-(1→6) linkages, when using saccharides as acceptors [80, 90, 91]. Indeed, *B. circulans* β-galactosidase was used for the synthesis of N-acetyllactosamine using *para*-nitrophenyl-β-D-galactopyranoside (PNPG) as donor and *N*-acetylglucosamine as acceptor, producing the β-(1→4) and β-(1→6) isomers in a (90:10) rate. However when the enzyme was used immobilized onto Eupergit C it was regioselective for the formation of β-(1→4) linkages [90]. Recently it has been reported that the use of co-solvents derived from glycerol in the reaction medium is able to direct the regioselectivity of the transglycosylation reaction, allowing to obtain only the β-(1→6) regioisomer when using *B. circulans* β-galactosidase [88].

6.4. Inhibition of the Transgalactosylation Synthesis

A point to be considered when working with the transgalactosylation process is that both, substrates and products, can be β-galactosidase inhibitors. One of the most common type of inhibition, as previously mentioned, is that exerted by the reaction products. In fact lactose hydrolysis products, glucose and galactose, proved to be inhibitors of *K. lactis* β-galactosidase, while galactose is inhibitor of *A. oryzae* β-galactosidase [44, 92, 93]. Kim et al. (2004) reported for the β-galactosidase from *K. lactis*, that galactose acts like a competitive inhibitor at low lactose concentration (below 50 mM) with a inhibition constant (K_i)

of 90 mM. Yet, at higher lactose concentrations, where lactose can also function as acceptor and transgalactosylation reaction takes place, galactose does not behave as an inhibitor as it takes part in the transgalactosylation reaction [94].

Several molecules which have been successfully used as acceptors in the transgalactosylation system catalyzed by β-galactosidase, have proved to be β-galactosidase inhibitors. *K. lactis* and *A. oryzae* enzymes were inhibited by moderate concentrations of allyl alcohol; whereas the *A. oryzae* β-galactosidase was inhibited by ethanol at concentrations higher than 4 M [63]. Methyl-β-D-xylopyranoside and benzyl-β-D-xylopyranoside showed to be uncompetitive and mixed inhibitors of *E. coli* β-galactosidase respectively [70].

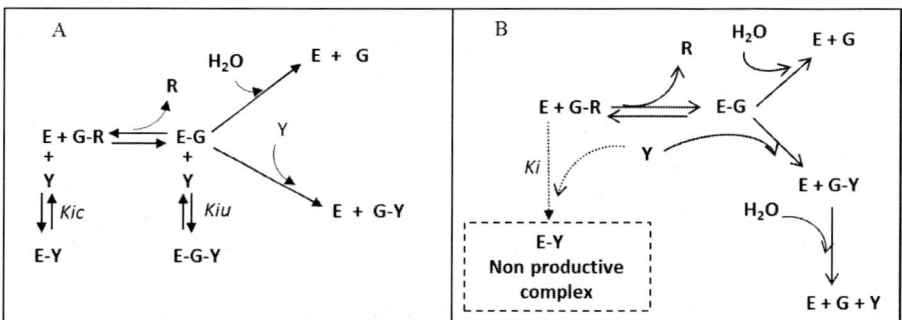

Figure 8. A) Mixed inhibition. B) Substrate like inhibition for substituted enzyme mechanism E: Enzyme; G-R: Galactose donor, G: Galactose; Y: acceptor molecule.

2-Aminoethanol proved to be inhibitor of *A. oryzae* β-galactosidase, fitting to a mixed inhibition model. This inhibition mechanism predicts binding of the inhibitor to the free enzyme giving an enzyme-inhibitor complex with a dissociation constant Kic, and to the galactosyl-enzyme complex giving an enzyme-galactose- inhibitor complex with a dissociation constant Kiu (Figure 8A). The Kic and Kiu determined were 0.310 ± 0.040 M and 0.604 ± 0.035 M respectively, using ONPG as substrate [74].

High concentrations of ethyleneglycol and glycerol exert substrate like inhibition of *A. oryzae* β-galactosidase, which can take place in enzymes that fits to substituted enzyme mechanism. This sort of inhibition happens when the acceptor binds to the free enzyme leading to a dead-end complex (Figure 8B). Ki values were 244 mM and 618 mM for glycerol and ethyleneglycol respectively [71, 95].

7. APPLICATIONS OF β-GALACTOSIDASES

7.1. Lactase in Dairy Industry

Lactose hydrolysis is one of the most developed enzymatic applications in dairy industry. As already described above, lactases are important in reducing lactose in milk products for lactose-intolerant people in order to protect them from severe diarrhea, fatal consequences, and tissue dehydration. However, there are other reasons besides those for which the hydrolysis of lactose in milk and its by-products is very beneficial, such as to enhance creaminess, sweetness, tastiness, and digestibility, and to decrease sandiness because of crystallization that occurs in lactose concentrated preparations, such as ice cream, yogurt, frozen dessert and condensed milk [96]. Cheese produced from hydrolyzed milk ripens more quickly compared to cheese produced from normal milk. The products of lactose hydrolysis (i.e., glucose and galactose) ferment more easily, thus reducing the overall time required to achieve the preferred pH in various food items such as yogurts and cottage cheese. Furthermore, it also reduces the need to add additional sweeteners, thus lowering the amount of calories in the final product [97].

Lactose and protein are the major components in whey and account for approximately 75 and 10% of the total solid, respectively [98]. Large amounts of whey are produced annually because approximately 9 kg of whey is obtained per kilogram of cheese produced. Whey protein can be extracted from whey by ultrafiltration and used for other purposes. Nevertheless, the remaining liquid, called whey permeate, which is composed mainly of lactose, salts, non-protein nitrogen and water [99], is currently discarded as a dairy effluent. From an environmental point of view, lactose is associated with the high biochemical and chemical oxygen demand (BOD/COD) content of whey. With the ever-increasing market for whey protein concentrates (WPC), a lactose-rich stream is produced (permeate) with reduced economical value, but still with a high pollutant load [100]. Several alternatives exist for cheese whey/permeate treatment and β-galactosidase application certainly is one of these. The enzyme can be employed directly on the cheese whey/permeate stream resulting in high sweetness syrups used as additive in ice creams, desserts, etc. Another possibility for the valorization of cheese whey/permeate is the hydrolysis of lactose by β-galactosidase and then the use of the hydrolyzed whey/permeate as a substrate for the production of bioethanol by microorganisms. Subsequent treatment of this organic waste with β-galactosidase can convert it into readily available substrate for cell cultivation [101].

Dairy yeasts, like *K. lactis* and *K. fragillis,* with a pH optimum 6.5–7.0 are generally used for the hydrolysis of lactose in milk or sweet whey. On the other hand, the fungal β-galactosidases, like those from *A. oryzae* and *A. niger,* with optimum pH 3–5 are more suitable for acidic whey hydrolysis [42, 43].

Despite of the several advantages of the enzymatic process compared to the use of traditional catalysts, there are a few practical problems associated with their employment in industrial applications. Enzymes are generally expensive, which means that the cost of their isolation and purification is many times higher than that of ordinary catalysts. Since they are structurally proteins, they are also highly sensitive to various denaturing conditions when isolated from their natural environments such as temperature, pH, and substances at trace levels, can act as inhibitors. Unlike conventional heterogeneous chemical catalysts, most enzymes work dissolved in water in homogeneous catalysis systems, leading to contamination of the product, which hinders the recovery of the enzyme and its use [102]. One of the most successful methods proposed to overcome these limitations is the use of immobilization strategies [103]. Immobilization is a technological process in which enzymes are fixed to or within solid supports, creating an heterogeneous immobilized enzyme system. The use of immobilized enzymes systems allows the easy recovery of both enzymes and products, multiple reuse of enzymes, continuous operation of enzymatic processes, rapid ending of enzymatic reactions, making the enzyme-based processes cost-effective and industrially viable [104,105].

β-Galactosidases have been immobilized by several methods such as entrapment, cross-linking, adsorption, covalent binding, or a combination of methods [106]. Many materials has been used as solid carriers for enzyme immobilization, including natural, synthetic, or magnetic carriers. These immobilized enzyme systems have also been applied in batch, fed-batch, and continuous production. Since each method has its own advantages and drawbacks, the selection of a suitable immobilization method depends on the enzyme (the different properties of various β-galactosidases, such as molecular weight, protein chain length, and position of the active site), matrix, reaction conditions, reactor, etc. [107, 108].

7.2. β-Galactosidase with Synthetic Purposes

7.2.1. Synthesis of Galactooligosaccharides

Galactooligosaccharides (GOS) are polysaccharides composed by several galactose moieties. They can be synthesized by transgalactosylation from lactose or lactulose catalyzed by β-galactosidases. Lactose or lactulose fulfils both, the roles of galactose donor and acceptor and the GOS formed can contain a glucose or fructose moiety at their reducing end respectively. GOS are generally termed prebiotics as they are non-digestible carbohydrates that stimulates the proliferation of bifidobacteria and lactobacillus in the intestine at the expense of less desirable organisms. Health benefits of prebiotics includes lactose tolerance, anti-tumorigenic activity, and formation of short chain fatty acids among others. In this context, GOS are widely used as functional foods additives [72, 109, 110].

Parallel to GOS formation, lactose hydrolysis can take place, and the prevalence of each reaction depends on lactose concentration. At low lactose concentrations, hydrolysis reaction mainly takes place while GOS formation is increased at high lactose concentrations. Neri et al. (2009) [109] reported GOS (mainly trisaccharides) reaction yields of 39.5% and 64.1% with initial lactose concentrations of 5% v/v and 50% v/v respectively, using *A. oryzae* β-galactosidase. In addition, yield and composition of GOS is highly dependent on the enzyme source. Formation of tri and tetrasaccharides (GOS-3 and GOS-4 respectively) is mainly observed for *B. circulans* and *A. oryzae* β-galactosidases while di and trisaccharides (GOS-2 and GOS-3 respectively) synthesis is reported for *K. lactis* β-galactosidase. β-(1→6) linkages are mainly formed by *A. oryzae* β-galactosidase, followed by β-(1→3) and β-(1→4), while *K. lactis* enzyme forms β-(1→6) linkages. Synthesis of the disaccharides β-D-galactopyranosyl-(1→6)-D-galactopyranose and β-D-galactopyranosyl-(1→6)-D-glucopyranose (allolactose) as well as the trisaccharide β-D-galactopyranosyl-(1→6)-β-D-galactopyranosyl-(1→4)-D-glucopyranose have been reported for *K. lactis* β-galactosidase. On the other hand, GOS synthesized mainly by *B. circulans* β-galactosidase are the trisaccharide β-D-galactopyranosyl-(1→4)-β-D-galactopyranosyl-(1→4)-D-glucopyranose and the tetrasaccharide β-D-galactopyranosyl-(1→4)-β-D-galactopyranosyl-(1→4)-β-D-galactopyranosyl-(1→4)-D-glucopyranose,
confirming the preference for the formation of β-(1→4) linkages. Other reported compounds for this enzyme source are the disaccharides β-D-galactopyranosyl-(1→6)-D-glucopyranose (allolactose), β-D-galactopyranosyl-(1→4)-D-galactopyranose (galactobiose), β-D-galactopyranosyl-(1→3)-D-glucopyranose and the trisaccharides β-D-galactopyranosyl-(1→6)-β-D-galactopyranosyl-(1→4)-D-glucopyranose and β-D-galactopyranosyl-(1→4)-β-D-galactopyranosyl-(1→3)-D-glucopyranose [111, 112].

When lactulose (β-D-galactopyranosyl-(1→4)-D-fructofuranose) is used both as galactose donor and acceptor by *K. lactis* β-galactosidase, two GOS-3 are formed. A mixture of the isomers β-D-galactopyranosyl-

(1→6)-β-D-galactopyranosyl-(1→4)-β-D-fructopyranose (the main one), β-D-galactopyranosyl-(1→6)-β-D-galactopyranosyl-(1→4)-β-D-fructofuranose and β-D-galactopyranosyl-(1→6)-β-D-galactopyranosyl-(1→4)-α-D-fructofuranose. The second GOS-3 formed is the mixture of the regioisomers β-D-galactopyranosyl-(1→4)-β-D-fructopyranosyl-(1→1)-β-D-galactopyranose (the main one) and β-D-galactopyranosyl-(1→4)-β-D-fructofuranosyl-(1→1)-β-D-galactopyranose and β-D-galactopyranosyl-(1→4)-α-D-fructofuranosyl-(1→1)-β-D-galactopyranose. When β-galactosidase from *A. oryzae* is used, the synthesis of the following compounds is achieved: β-D-galactopyranosyl-(1→6)-D-galactopyranose (6-galactobiose), β-D-galactopyranosyl-(1→6)-D-fructofuranose (allolactulose) and β-D-galactopyranosyl-(1→6)-β-galactopyranosyl (1→4)-β-D-fructofuranose (mayor compound produced) [105, 113, 114].

Among GOS-2, lactulose (β-D-galactopyranosyl-(1→4)-D-fructofuranose) is the most studied. It has been used for the treatment of constipation and some liver diseases in humans as well as in the food industry as a functional ingredient due to its prebiotic properties. Commercial lactulose is mainly produced through chemical isomerization of lactose under alkaline media. However, the enzymatic synthesis is an alternative to the chemical synthesis drawbacks such as removal of the catalysts and by products. β-galactosidases from different sources has been used to synthesize lactulose using lactose as galactose donor and fructose as acceptor. In this case, different products are generated depending on the enzyme source and the lactose/fructose initial molar ratio. *A. oryzae* β-galactosidase produces both lactulose and GOS (mainly GOS-3 and GOS-4), β-galactosidase from *K. lactis* produces mainly GOS-2 and lactulose to a lesser extent, while the enzyme from *B. circulans* produces mainly GOS-3 with small amounts of lactulose. On the other hand *K. lactis* β-galactosidase produces β-D-galactopyranosyl-(1→1)-β-D-fructofuranose (1-lactulose) in addition to lactulose [115, 116].

7.2.2. Synthesis of Alkylgalactosides

Alkylglycosides are a group of non-ionic, biodegradable, hypoallergenic and chemical stable surface active agents, with several potential applications in pharmaceutical, food and cosmetic industries. They contain a carbohydrate hydrophilic head (galactose) and a hydrophobic hydrocarbon tail, usually derived from a primary fatty alcohol. They can be enzymatically synthesized by transfer of a galactose moiety from a galactose donor to an aliphatic alcohol [117].

Lipid coated β-galactosidases from *E. coli*, *A. oryzae* and *B. circulans* were used for the synthesis of alkylgalactosides in dry diisopropylether using PNPG as the galactose donor. While high reaction yields were achieved for aliphatic primary alcohol with short chain length (less than 8 C), the reactivity decreases for long alcohol chains. Lipid coated β-galactosidase from *E. coli* and *A. oryzae* showed preference for primary alcohols while that of *B. circulans* showed preference for secondary alcohols [80].

The synthesis of propyl-β-D-galactoside was achieved with immobilized *A. oryzae* β-galactosidase, using lactose as the galactose donor; even though the enzyme showed a low stability in 85% 1-propanol ($t_{1/2}$ = 3.1 h) [117]. Butyl-β-D-galactoside could also be synthesized using *A. oryzae* β-galactosidase, either with lactose or ONPG as the galactose donor and 1-butanol as acceptor. When using ONPG as donor, yields of 76% were achieved. The nature of the galactose donor was critical in the product yield, being lower for lactose than for ONPG. Nevertheless, under optimized conditions reasonable yields (58%) were obtained with lactose, a cheaper donor. β-galactosidases from other sources failed to synthesized butyl-galactoside. Very low amounts could be produced with *B. circulans* β-galactosidase (6%), while none at all with the enzyme from *K. lactis*. Unlike *A. oryzae* β-galactosidase, *K. lactis* β-galactosidase is unstable in 1-butanol ($t_{1/2}$ lower than 3 min) which explains the poor results obtained with this enzyme in the reaction synthesis [118, 119].

The synthesis of hexyl-β-D-galactoside was feasible using methyl-β-D-galactoside and hexanol as donor and acceptor substrates respectively with *A. oryzae* β-galactosidase. High reaction yields were achieved in acetone: water (30:70) medium. The leaving group of the donor substrate was a relevant parameter obtaining reaction yields of 6%, 12%, 59%, 47% and 49% when lactose, lactulose, ONPG, propyl-β-D-galactoside or butyl-β-D-galactoside were used as donor substrates respectively [120] [121] [122].

In addition, *O*-allyl-β-D-galactopyranoside was synthesized using *A. oryzae* β-galactosidase. The reaction was carried out employing PNPG as the donor and a large excess of allyl alcohol as the acceptor with a molar yield of 65.6% [123]. Among other, alkylgalactosides synthesized with *K. lactis* β-galactosidase, using lactose as galactose donor, hexanediol-galactoside (6-hydroxyhexyl-β-D-Galactopyranoside), butanediol-galactoside (4-hydroxybutyl-β-D-galactopyranoside), glycerol-galactoside (glyceryl-β-D-galactopyranoside), benzyl-galactoside (benzyl-β-D-galactopyranoside) and ethyl-galactoside (ethyl-β-D-galactopyranoside) as acceptors were reported. Reaction yields were among 21-60% [69].

7.2.3. Synthesis of Galactosides with Potential Biological Activity

Galactosylation of bioactive compounds can lead to changes in the activity of the parent molecules and can also enhance its pharmacokinetic parameters, such as solubility and hydrophilicity, which facilitate their passage through the cell membrane. Several examples of galactosylation of hydroxylated molecules have been reported in the literature. Hydroquinone has been galactosylated forming 4-hydroxy-phenyl-β-D-galactopyranoside (hydroquinone-galactoside), even though with low yields. The galactosylation was assayed with β-galactosidases from different sources (*K. lactis*, *A. oryzae*, *B. cirulans* and *Thermus sp*) and the highest amount of galactoside was obtained with *K. lactis* β-galactosidase, 2.29 mM of 4-hydroxy-phenyl-β-D-galactopyranoside starting from 100 mM lactose and 100 mM hydroquinone [82].

A β-galactosidase isolated from an anaerobic extreme thermophile, *Thermoanaerobacter sp* was successfully used for the enzymatic galactosylation of 1D-*chiro*-inositol, 1D-pinitol, 1D-3-*O*-allyl-4-*O*-methyl-

chiro-inositol, 1D-3,4-di-O-methyl-*chiro*-inositol, 1L-*chiro*-inositol and *myo*-inositol, with yields ranging from 46% to 64%, and using PNPG as the galactose donor [124].

On the other hand, chlorphenesin (CPN) is used as a muscle relaxant as CPN carbamate, particularly to treat injuries and other painful muscular conditions. It is also widely used as a preservative in cosmetics. Its galactosylation can increase it solubility, decrease its toxicity and improve targeting capability. Galactosylation was performed using *E. coli* β-galactosidase starting from lactose (400 g/L) and CPN (44 mM). Maximum reaction yield (64%) was achieved upon 12 hours. Galactosylated CPN showed similar minimal inhibitor concentration to that of CPN against microorganisms and was notoriously less cytotoxic against HACAT cells [125].

The synthesis of 2-fluoroethyl-glycosides as "protoxin" forms of sodium monofluoroacetate, for use in the control of introduced mammalian pest species, has been reported by Stevenson et al. (1996). A major drawback of the used toxins is that they can be tasted or smelled by some animals, which in turn would avoid their ingestion. Therefore, the generation of non-volatile, hydrolytically stable protoxins, which can be easily converted to free toxin in the digestive system of the target species, could be an interesting alternative. The synthesis of 2-fluoroethyl-β-D-galactopyranoside was achieved with a β-galactosidase from *Streptococcus thermophilus* with reaction yields of 40%. Lactose and fluoroethanol were used as donor and acceptor respectively [126].

Other interesting compounds such as galactose containing disaccharide nucleosides has been synthesized by using the transgalactosylation potential of β-galactosidase from *A. oryzae*. PNPG was used as the galactose donor and 2-deoxyuridine, uridine, thymidine and adenosine as the acceptors. The products obtained were β-D-galactopyranosyl-(1→5') 2'-deoxyuridine (7%), β-D-galactopyranosyl (1→5') uridine (6%), β-D-galactopyranosyl (1→5') thymidine and β-D-galactopyranosyl (1→3') thymidine (2:1) with 5% yield, β-D-galactopyranosyl (1→5') adenosine and β-D-galactopyranosyl (1→3') adenosine (3.3:1) with a yield of 3% [127].

β-galactosidase was also useful for the generation of nucleotide-activated oligosaccharides. The synthesis was performed using *B. circulans* β-galactosidase, lactose as donor and the nucleotide sugars UDP-GlcNAc and UDP-Glc as acceptors. The following products were obtained stereo and regioselectively: β-D-galactopyranosyl-(1→4)α-D-N-acetylglucopyranosyl-1-uridine-diphosphate (UDPLacNAc), β-D-galactopyranosyl-(1→4)-α-D-glucopyranosyl 1-uridine-diphosphate (UDP-Lac) and the UDP-trisaccharides β-D-galactopyranosyl-(1→4)-β-D-galactopyranosyl-β-(1→4)-α-D-N-acetylglucopyranosyl1-uridine-diphosphate and β-D-galactopyranosyl-(1→4)-β-D-galactopyranosyl-(1→4)- α-D-glucopyranosyl1-uridine-diphosphate

The analysis of other nucleotide sugars revealed UDP-Gal, UDP-GalNAc, UDP-Xyl and dTDP-, CDP-, ADP- and GDP-Glc as further acceptor substrates for β–galactosidase from *B. circulans* [128]. In this context, an access to nucleotide-activated oligosaccharides by chemical and/or enzymatic methods could help to elucidate the biosynthesis and biological function of these naturally occurring glyco-conjugates.

Another application for the β-galactosidase is the galactosylation of antibiotics. Chlorphenisin and chloramphenicol were galactosylated by *A. oryzae* β-galactosidase using lactose as the galactose donor. Reaction yields were improved by the use of 20% (v/v) acetonitrile as cosolvent. Chlorphenisin and chloramphenicol were selectively galactosylated at their primary hydroxyl groups [129].

The enzymatic synthesis of the disaccharide β-D-galactopyranosyl-(1→4)-D-xylose is interesting as it can be used to evaluate the in vivo activity of intestinal β-galactosidase via a non-invasive method. In this context the synthesis of 4-, 3-, and 2-o-β-D-galactopyranosyl-D-xylosides could be achieved by enzymatic transgalactosylation using both, ONPG and lactose as galactosyl donor, and xylose as acceptor using β-galactosidase from *E. coli*. Morevoer, *E. coli* β-galactosidase showed specificity for the generation of β-(1→4) and β-(1→3) glycosidic linkages when using substituted xylopyranose as acceptor and ONPG as galactosyl donor. The enzyme was regioselective for the synthesis of β-D-galactopyranosyl-(1→4)-D methylxylopyranoside while a mixture of β-D-galactopyranosyl-(1→3)-D benzylxylopyranoside and β-D-galactopyranosyl-(1→4)-D benzylxylopyranoside (84:16) was achieved [70, 129].

Due to its excellent transgalactosylation properties A. oryzae β-galactosidase has been used for the generation of diverse galactosides with potential activity as galectin inhibitors. Galectins are a family of carbohydrate-recognizing proteins that can interact with specific glycoepitopes and mediate important biological processes, including immune cell homeostasis and activation of tolerogenic circuits. Particularly, Galectin 1 and 3 have shown pro-tumorigenic effects, being overexpressed in numerous tumor cells, contributing to tumor immune escape, tumor progression and resistance to drug-induced apoptosis. Thus, the generation of specific glycosides with known structure, that could compete with their natural ligands blocking galectin carbohydrate recognition domain (CRD) is one of the current major challenges in the field, since these galactosides could be potential antitumor agents. In this context our research group has synthesized several galactosides using A. oryzae β-galactosidase and either lactose or ONPG, as galactose donor. We achieved the synthesis of 2-O-β-D-galactopyranosyl- ethylene glycol, a (1:1) mixture of (2R) and (2S)-3-O-β-D-galactopyranosyl- glycerol, and a (1:1) mixture of (2R,3S) and (2S,3R)-4-O-β-D-galactopyranosyl-erythritol using ethyleneglycol, glycerol and erithrytol as acceptors respectively. We also used aminoalcohols as acceptors generating 2-aminoethyl-β-D-galactopyranoside and 3-aminopropyl-1-O-β-D- galactopyranoside. Finally the synthesis of non-natural trisaccharides was assayed obtaining β-D-galactopyranosyl-(1→6)- β-D-galactopyranosyl-(1→4)-D-Glucopyranose and a mixture of the regioisomers β-D-Galactopyranosyl-(1→6)- β-D-Glucopyranose-(1→4)-D-Glucopyranose and β-D-Galactopyranosyl-(1→3)- β-D-Glucopyranosyl-(1→4)-D-Glucopyranose. Unfortunately, none of these compounds proved to be better ligands than natural galectin ligands. Yet some of these galactosides constitutes functionalized galactose that are interesting molecules for the generation of glyco-polymers and glycol-nanoparticles, for the development of biosensors, as well as for building blocks in chemical synthesis. Trisaccharides can also be used for the elucidation of the role of glycan-protein interactions in biological systems [73, 74, 84, 89]. In the search of galectin inhibitors the enzymatic synthesis of galactosides of salicin was performed using *A. oryzae* β-galactosidase [131].

7.3. β-Galactosidases as Analytical Tools in Glyco Biology

Glycans present in biological glycol-conjugates have several structural and functional roles. In fact, changes in glycan profile of cell glycoproteins and glycolipids are associated with the development of several diseases. Moreover, glycans from different pathogens are involved both in pathogen infection as well as in their immune evasion strategies. In this context, the elucidation of glycan structure as well as enlightening of their biological function is essential to understand their role in pathogenesis.

Exoglycosidases, such as β-galactosidases are powerful tools used for glycan analysis due to their high specificity for the removal of carbohydrates from glycoproteins without altering the protein structure. In addition, the use of immobilized enzymes allows their easy removal from the reaction mixture avoiding further purification steps. The subsequent evaluation of the changes generated in the biological function of glycoproteins due to specific sugar removal contributes to a better understanding of their role in the process involved. In this sense, the knowledge of the enzyme specificity is essential for their use as analytical tools [131, 132].

Dragosits et al. (2014) [93] evaluated the specificity of two previously uncharacterized β-galactosidases from *A. nidulans*, as well as the *A. oryzae* β-galactosidase, assaying several linear and branched galactooligosaccharides with different glycosidic linkages as substrates. None of the reported β-galactosidases up to this research showed an absolute specificity toward either terminal β-(1, 3)-linked, β-(1, 4) - linked, or β-(1, 6)-linked galactose. *E. coli*, *A. oryzae*, *Canavalia ensiformis* (jack beans), and *S. pneumoniae* mainly cleaves terminal β- (1, 4)-linked galactose and to a lesser extent β-(1, 3)-linked galactose. On the other hand, bovine testes β- galactosidase cleaves terminal β- (1, 3) -linked galactose, but can also cleave β-(1,4)-linked galactose efficiently. Nevertheless, *A. nidulans* rlacA β-galactosidase described by Dragosit et al. in 2014, showed preference toward β-(1,3)-linked galactose as well as the ability to remove β-(1,3)-linked galactose residues from the dabsylated-GalGal glycopeptide. *C. ensiformis* β-galactosidase cannot release galactose β-(1,4)-linked to a core fucose residue, while *S. pneumoniae* β-galactosidase could digest that epitope. The *A. nidulans* rlacA β-galactosidase is able to remove galactose β-(1,4)-linked to a core fucose residue [93]. Rodriguez et al. (2018) described the immobilization of *A. oryzae* β-galactosidase and further applications to the selective degalactosylation of model proteins such as asialofetuin and complex protein samples such as *Fasciola hepatica* lysate, which contains mixture of proteins as well as other cellular components. A 50% decrease in peanut (Arachis hypogaeae) lectin (PNA lectin) recognition was observed due to selective enzymatic treatment of asialofetuin with immobilized β-galactosidase. This lectin is specific for recognition of terminal galactose residues. In the same way a decrease of 55% of PNA lectin recognition on the *F. hepatica* lysate treated with β-galactosidase was observed [133]. Iskratch et al. (2009) also described the use of β-galactosidases from bovine testes and *A. oryzae* for the generation of asialofetuin with different glycomic pattern for the analysis of lectin specificities [134].

REFERENCES

[1] Henrissat, B. A classification of glycosyl hydrolases based on amino acid sequence similarities. *Biochem. J.* 1991, 280, 309–316.

[2] Henrissat, B.; Bairoch, A. New families in the classification of glycosyl hydrolases based on amino acid sequence similarities. *Biochem. J.* 1993, 293, 781–788.

[3] Henrissat, B.; Bairoch, A. Updating the sequence-based classification of glycosyl hydrolases. *Biochem. J. Lett.* 1996, 316, 695–696.

[4] Lo, J.; Mukerji, K.; Awasthi, Y.C.; Hanada, E.; Suzuki, K.; Srivastava, S.K. Purification and properties of sphingolipid ß-galactosidases from human placenta. *J. Biol. Chem.* 1979, 254, 6710–6715.

[5] Bhalla, P.L.; Dalling, M.J. Characteristics of a ß-galactosidase associated with the stroma of chloroplasts prepared from mesophyll protoplasts of the primary leaf of wheat. *Plant Physiol.* 1984, 76, 92–95.

[6] Maley, F.; Trimble, R.B.; Tarentino, A.L.; Plummer, T.H. Characterization of glycoproteins and their associated oligosaccharides through the use of endoglycosidases. *Anal. Biochem.* 1989, 180, 195–204.

[7] Raghothama, K.G.; Lawton, K.A.; Goldsbrough, P.B.; Woodson, W.R. Characterization of an ethylene-regulated flower senescence-related gene from carnation. *Plant Mol. Biol.* 1991, 17, 61–71.

[8] De Veau, E.J.I.; Gross, K.C.; Huber, D.J.; Watada, A.E. Degradation and solubilization of pectin by ß-galactosidases purified from avocado mesocarp. *Physiol. Plant.* 1993, 87, 279–285.

[9] Ross, G.S.; Redgwell, R.J.; MacRae, E.A. Kiwifruit ß-galactosidase: Isolation and activity against specific fruit cell-wall polysaccharides. *Planta* 1993, 189, 499–506.

[10] Buckeridge, M.S; Grant Reid, J.S. Purification and properties of a novel ß-galactosidase or exo-(1-4)-ß-D-galactanase from the cotyledons of germinated *Lupinus angustifolius* L. seeds. *Planta* 1994, 192, 502–511.

[11] Hall, B.G.; Malik, H.S. Determining the evolutionary potential of a gene. *Mol. Biol. Evol.* 1998, 15, 1055–1061.

[12] Ingram, C.J.E.; Mulcare, C.A.; Itan, Y.; Thomas, M.G.; Swallow, D.M. Lactose digestion and the evolutionary genetics of lactase persistence. *Hum. Genet.* 2009, 124, 579–591.

[13] O'Connell, S.; Walsh, G. Application relevant studies of fungal ß-galactosidases with potential application in the alleviation of lactose intolerance. *Appl. Biochem. Biotechnol.* 2008, 149, 129–138.

[14] Suarez, F.L.; Savaiano, D.; Arbisi, P.; Levitt, M.D. Tolerance to the daily ingestion of two cups of milk by individuals claiming lactose intolerance. *Am. J. Clin. Nutr.* 1997, 65, 1502–1506.

[15] Hertzler S.R.; Savaiano, D.A.; Levitt, M.D. Fecal hydrogen production and consumption measurements. *Dig. Dis. Sci.* 1997, 42, 348–353.

[16] Saltzman, J.R.; Russell, R.M.; Golner, B.; Barakat, S.; Dallal, G.E.; Goldin, B.R. A randomized trial of *Lactobacillus acidophilus* BG2FO4 to treat lactose intolerance. *Am. J. Clin. Nutr.* 1999, 69, 140–146.

[17] Jones, B.L.; Raga, T.O.; Liebert, A.; Zmarz, P.; Bekele, E.; Danielsen, E.T.; Olsen, A.K.; Bradman, N.; Troelsen, J.T.; Swallow, D.M. Diversity of lactase persistence alleles in Ethiopia: Signature of a soft selective sweep. *Am. J. Hum. Genet.* 2013, 93, 538–544.

[18] Kobayashi, T.; Shinnoh, N.; Goto, I.; Kuroiwa, Y. Hydrolysis of galactosylceramide is catalyzed by two genetically distinct acid ß-galactosidases. *J. Biol. Chem.* 1985, 260, 14982–14987.

[19] Wallenfels, K.; Malhotra, O.P. Galactosidases. *Adv. Carbohydr. Chem.* 1961, 16, 239–298.

[20] de Alcântara, P.H.N.; Martim, L.; Silva, C.O.; Dietrich, S.M.C.; Buckeridge, M.S. Purification of a ß-galactosidase from cotyledons of *Hymenaea courbaril* L. (Leguminosae). Enzyme properties and biological function. *Plant Physiol. Biochem.* 2006, 44, 619–627.

[21] Seddigh, S.; Darabi, M. Comprehensive analysis of ß-galactosidase protein in plants based on *Arabidopsis thaliana*. *Turkish J. Biol.* 2014, 38, 140–150.

[22] Kim, J.; Gross, K.C.; Solomos, T. Galactose metabolism and ethylene production during development and ripening of tomato fruit. *Postharvest Biol. Technol.* 1991, 1, 67–80.

[23] Smith, D.L.; Gross, K.C. A family of at least seven ß-galactosidase genes is expressed during tomato fruit development. *Plant Physiol.* 2000, 123, 1173–1183.

[24] Hirano, Y.; Tsumuraya, Y.; Hashimoto, Y. Characterization of spinach leaf a-L-arabinofuranosidases and ß-galactosidases and their synergistic action on an endogenous arabinogalactan-protein. *Physiol. Plant.* 1994, 92, 286–-296.

[25] Li, S.C.; Han, J.W.; Chen, K.C.; Chen, C.S. Purification and characterization of isoforms of ß-galactosidases in mung bean seedlings. *Phytochemistry* 2001, 57, 349–359.

[26] Kotake, T.; Dina, S.; Konishi, T.; Kaneko, S.; Igarashi, K.; Samejima, M.; Watanabe, Y.; Kimura, K.; Tsumuraya, Y. Molecular cloning of a ß-galactosidase from radish that residues of arabinogalactan Protein 1. *Plant Physiol.* 2005, 138, 1563–1576.

[27] Wu, A.; Liu, J. Isolation of the promoter of a cotton ß-galactosidase gene (GhGal1) and its expression in transgenic tobacco plants. *Sci. China, Ser. C Life Sci.* 2006, 49, 105–114.

[28] Rimlumduan, T.; Hua, Y. ling; Tanaka, T.; Ketudat Cairns, J.R. Structure of a plant ß-galactosidase C-terminal domain. *Biochim. Biophys. Acta.* 2016, 1864, 1411–1418.

[29] Mlichová, Z.; Rosenberg, M. Current trends of ß-galactosidase application in food technology. *J. Food Nutr. Res.* 2006, 45, 47–54.

[30] Rubio-Texeira, M. Endless versatility in the biotechnological applications of Kluyveromyces LAC genes. *Biotechnol. Adv.* 2006, 24, 212–225.

[31] Karlsson, U.; Koorajian, S.; Zabin, I.; Sjöstrand, F.S.; Miller, A. High resolution electron microscopy on highly purified ß-galactosidase from *Escherichia coli*. *J. Ultrasructure Res.* 1964, 10, 457–469.

[32] Jacobson, R.H.; Matthews, B.W. Crystallization of ß-galactosidase from *Escherichia coli*. *J. Mol. Biol.* 1992, 223, 1177–1182.

[33] Fowler, A. V.; Zabin, I. The amino acid sequence of ß-galactosidase of *Escherichia coli. Proc. Natl. Acad. Sci.* 1977, 74, 1507–1510.
[34] Kalnins, A.; Otto, K.; Rüther, U.; Müller-Hill, B. Sequence of the lacZ gene of *Escherichia coli. Embo J.* 1983, 2, 593–597.
[35] Matthews, B.W. The structure of *E. coli* ß-galactosidase. *C. R. Biol.* 2005, 328, 549–556.
[36] Humphrey, W.; Dalke, A.; Schulten, K. VMD-Visual Molecular Dynamics. *J. Mol. Graph.* 1996, 14, 33–38.
[37] Juers, D.H.; Rob, B.; Dugdale, M.L.; Rahimzadeh, N.; Giang, C.; Lee, M.; Matthews, B.W.; Huber, R.E. Direct and indirect roles of His-418 in metal binding and in the activity of ß-galactosidase (*E. coli*). *Protein Sci.* 2009, 18, 1281–1292.
[38] Ito, Y.; Sasaki, T.; Kitamoto, K.; Kumagai, C.; Takahashi, K.; Gomi, K.; Tamura, G. Cloning, nucleotide sequencing, and expression of the ?-galactosidase-encoding gene (lacA) from *Aspergillus oryzae. J. Gen. Appl. Microbiol.* 2002, 48, 135–142.
[39] Maksimainen, M.M.; Lampio, A.; Mertanen, M.; Turunen, O.; Rouvinen, J. The crystal structure of acidic ß-galactosidase from *Aspergillus oryzae. Int. J. Biol. Macromol.* 2013, 60, 109–115.
[40] Pereira-Rodríguez, Á.; Fernández-Leiro, R.; González-Siso, M.I.; Cerdán, M.E.; Becerra, M.; Sanz-Aparicio, J. Structural basis of specificity in tetrameric *Kluyveromyces lactis* ß-galactosidase. *J. Struct. Biol.* 2012, 177, 392–401.
[41] Becerra Fernandez, M.; Cerdan Villanueva, M.E.; González Siso, M.I. Micro-scale purification of ß-galactosidase from *Kluyveromyces lactis* reveals that dimeric and tetrameric forms are active. *Biotechnol. Tech.* 1998, 12, 253–256.
[42] Asraf, S.S.; Gunasekaran, P. Current trends in ß-galactosidase research and application. *Curr. Res. Technol. Educ. Top. Appl. Micobiology Microb. Biotechnol.* 2010, 2, 880.
[43] Lu, Li.; Xiao, M.; Li, Z.; Li, Y.; Wang, F. A novel transglycosylating ß-galactosidase from *Enterobacter cloacae* B5. *Process Biochem.* 2009, 44, 232–236.

[44] Cavaille, D.; Combes, D. Characterization of ß-galactosidase from *Kluyveromices lactis. Biotechnol. Appl. Biochem.* 1995, 22, 55–54.

[45] Alazzeh, A.Y.; Ibrahim, S.A.; Song, D.; Shahbazi, A.; AbuGhazaleh, A.A. Carbohydrate and protein sources influence the induction of a- and ß-galactosidases in *Lactobacillus reuteri. Food Chem.* 2009, 117, 654–659.

[46] Iqbal, S.; Nguyen, T.H.; Nguyen, T.T.; Maischberger, T.; Haltrich, D. ß-galactosidase from *Lactobacillus plantarum* WCFS1: Biochemical characterization and formation of prebiotic galacto-oligosaccharides. *Carbohydr. Res.* 2010, 345, 1408–1416.

[47] Anisha, G.S. ß-Galactosidases. *Curr. Dev. Biotechnol. Bioeng.* 2017, 395–421.

[48] Park, Y.K.; De Santi M.S.S.; Pastore, G.M. Production and characterization of ?-galactosidase from *Aspergillus oryzae. J. Food Sci.* 1979, 44, 100–103.

[49] Quinn Z.K, Z.; Xiao Dong, C. Effects of temperature and pH on the catalytic activity of the immobilized ß-galactosidase from *Kluyveromyces lactis. Biochem. Eng. J.* 2001, 9, 33–40.

[50] Roy, I.; Gupta, M.N. Lactose hydrolysis by Lactozym TM immobilized on cellulose beads in batch and fluidized bed modes. *Process Biochem.* 2003, 39, 325–332.

[51] Giacomini, C.; Irazoqui, G.; Batista-Viera, F.; Brena, B.M. Influence of the immobilization chemistry on the properties of immobilized ß-galactosidases. *J. Mol. Catal. - B Enzym.* 2001, 11, 597–606.

[52] Boon, M.A.; Janssen, A.E.M.; van't Riet, K. Effect of temperature and enzyme origin on the enzymatic. *Enzyme Microb. Technol.* 2000, 26, 271–281.

[53] Kuby, S. A.; Lardy, H.A. Purification and kinetics of ß-D-galactosidase from Escherichia coli, strain K-12. J. Am. Chem. Soc. 1953, 75, 890–896.

[54] Craven, R.; Anfiksen, B. Purification, composition and molecular weight ß-galactosidase of *Escherichia coli K12. J. Biol. Chem.* 1965, 240, 2468–2477.

[55] Premi, L.; Sandine, W.E.; Elliker, P.R.; De Klerk, H.C. Lactose-hydrolyzing enzymes of *Lactobacillus* Species. *Appled. Microbiol.* 1972, 24, 51–57.

[56] Singh, H.P.; Rao, M.V.R.; Dutta, S.M. Partial purification and properties of *Leuconostoc citrovorum* ?-galactosidase. *Milk Sci. Int.* 1979, 34, 475–478.

[57] Bahl, Om P. and Agrawal, K.M.L. Glycosidases of *Aspergillus niger*. *J. Biol. Chem.* 1969, 244, 2970–2978.

[58] Tanaka, Y.; Kagamiishi, A.; Kiuchi, A.; Horiuchi, T. Purification and properties of glutaminase from *Aspergillus oryzae*. *J. Biochem.* 1975, 77, 241–247.

[59] Rye, C.S.; Withers, S.G. Glycosidase mechanisms. *Curr. Opin. Chem. Biol.* 2000, 4, 573–580.

[60] Thuan, N.H.; Sohng, J.K. Recent biotechnological progress in enzymatic synthesis of glycosides. *J. Ind. Microbiol. Biotechnol.* 2013, 40, 1329–1356.

[61] Ichikawa, Y.; Look, G.C.; Wong, C.-H. Enzyme-catalyzed oligosaccharide synthesis. *Anal. Biochem.* 1992, 202, 215–238.

[62] Bojarová, P.; Kren, V. Glycosidases: a key to tailored carbohydrates. *Trends Biotechnol.* 2009, 27, 199–209.

[63] Van Rantwijk, F.; Woudenberg-Van Oosterom, M.; Sheldon, R.A. Glycosidase-catalysed synthesis of alkyl glycosides. *J. Mol. Catal. - B Enzym.* 1999, 6, 511–532.

[64] Palcic, M.M. Biocatalytic synthesis of oligosaccharides. *Curr. Opin. Biotechnol.* 1999, 10, 616–624.

[65] Scigelova, M.; Singh, S.; Crout, D.H.G. Glycosidases: a great synthetic tool. *J. Mol. Catal. B Enzym.* 1999, 6, 483–494.

[66] Juers, D.H.; Heightman, T.D.; Vasella, A.; McCarter, J.D.; Mackenzie, L.; Withers, S.G.; Matthews, B.W. A structural view of the action of *Escherichia coli* (lacZ) ß-galactosidase. *Biochemistry* 2001, 40, 14781–14794.

[67] Brás, P.A.; Fernandes, N.F..; Ramos, M.J. QM / MM Studies on the ß-galactosidase catalytic mechanism?: hydrolysis and transglycosylation reactions. *J. Chem. Theory Comput.* 2010, 6, 421–433.

[68] Crout, D.H.G.; Vic, G. Glycosidases and glycosyl transferases in glycoside and oligosaccharide synthesis. *Curr. Opin. Chem. Biol.* 1998, 2, 98–111.

[69] Stevenson, D.E.; Stanley, R.A.; Furneaux, R.H. Optimization of alkyl ß-D-galactopyranoside synthesis from lactose using commercially available ß-galactosidases. *Biotechnol. Bioeng.* 1993, 42, 657–666.

[70] López, R.; Fernández-Mayoralas, A. Enzymatic ß-galactosidation of modified monosaccharides: study of the enzyme selectivity for the acceptor and its application to the synthesis of disaccharides. *J. Org. Chem.* 1994, 59, 737–745.

[71] Irazoqui, G.; Bustamante, M.J.; Castilla, A.; Villagrán, L.V.; Batista-Viera, F.; Brena, B.M.; Giacomini, C. Substrate-like inhibition of the transgalactosylation reaction catalyzed by ß-galactosidase from *Aspergillus oryzae*. *Biocatal. Biotransformation* 2013, 31, 57–65.

[72] Maugard, T.; Gaunt, D.; Legoy, M.D.; Besson, T. Microwave-assisted synthesis of galacto-oligosaccharides from lactose with immobilized ß-galactosidase from *Kluyveromyces lactis*. *Biotechnol. Lett.* 2003, 25, 623–629.

[73] González, C.P.; Rodríguez, E.; Soule, S.; Fraguas, L.F.; Brena, B.M.; Giacomini, C.; Irazoqui, G. Enzymatic synthesis of 3-aminopropyl-1- O -ß- D -galactopyranoside catalyzed by *Aspergillus oryzae* ß-galactosidase. *Biocatal. Biotransformation.* 2015, 33, 197–207.

[74] Porciúncula González, C.; Castilla, A.; Garófalo, L.; Soule, S.; Irazoqui, G.; Giacomini, C. Enzymatic synthesis of 2-aminoethyl ß-d-galactopyranoside catalyzed by *Aspergillus oryzae* ß-galactosidase. *Carbohydr. Res.* 2013, 368, 104–110.

[75] Binder, W.H.; Kählig, H.; Schmid, W. Galactosylation by use of ß-galactosidase: Chemo-enzymatic syntheses of di- and trisaccharides. *Tetrahedron.* 1994, 50, 10407–10418.

[76] Crout, D.H.G.; MacManus, D.A.; Critchley, P. Enzymatic synthesis of glycosides using the ß-galactosidase of *Escherichia coli*: regio- and stereo-chemical studies. *J. Chem. Soc., Perkin Trans.* 1 1990, 0, 1865–1868.

[77] Usui, T.; Morimoto, S.; Hayakawa, Y.; Kawaguchi, M.; Murata, T.; Matahira, Y.; Nishida, Y. Regioselectivity of ß-D-galactosyl-disaccharide formation using the ß-D-galactosidase from *Bacillus circulans*. *Carbohydr. Res.* 1996, 285, 29–39.

[78] Zeng, X.; Yoshino, R.; Murata, T.; Ajisaka, K.; Usui, T. Regioselective synthesis of p-nitrophenyl glycosides of ß-D-galactopyranosyl-disaccharides by transglycosylation with ß-D-galactosidases. *Carbohydr. Res.* 2000, 325, 120–131.

[79] Vic, G.; Hastings, J.J.; Howarth, O.W.; Crout, D.H.G. Chemoenzymatic synthesis of ethyl 1-thio-(ß-D-galactopyranosyl)-0-ß-D-glycopyranosyl disaccharides using the ß-galactosidase from *Bacillus circulans*. *Tetrahedron Asymmetry* 1996, 7, 709–720.

[80] Okahata, Y.; Mori, T. Effective transgalactosylation catalysed by a lipid-coated ß-D-galactosidase in organic solvents. *J. Chem. Soc.* 1996, 0, 2861–2866.

[81] Lu, L.; Xu, L.; Guo, Y.; Zhang, D.; Qi, T.; Jin, L.; Gu, G.; Xu, L.; Xiao, M. Glycosylation of phenolic compounds by the site-mutated ß-galactosidase from *Lactobacillus bulgaricus* L3. *PLoS One* 2015, 10, 1–15.

[82] Kim, G.E.; Lee, J.H.; Jung, S.H.; Seo, E.S.; Jin, S. De; Kim, G.J.; Cha, J.; Kim, E.J.; Park, K.D.; Kim, D. Enzymatic synthesis and characterization of hydroquinone galactoside using *Kluyveromyces lactis* lactase. *J. Agric. Food Chem.* 2010, 58, 9492–9497.

[83] Potocká, E.; Mastihubová, M.; Mastihuba, V. Enzymatic synthesis of tyrosol glycosides. *J. Mol. Catal. B Enzym.* 2015, 113, 23–28.

[84] Porciúncula González, C.; Cagnoni, A.J.; Mariño, K. V.; Fontana, C.; Saenz-Méndez, P.; Irazoqui, G.; Giacomini, C. Enzymatic synthesis of non-natural trisaccharides and galactosides; Insights of their interaction with galectins as a function of their structure. *Carbohydr. Res.* 2019, 472, 1–15.

[85] Porciúncula González, D.C. *Development of transglycosylation systems enzymatic as a tool for the generation of bioactive compounds*. Masther's Thesis. http://riquim.fq.edu.uy/archive/files/683b9742fdca415cb608ffed5a1e1223.pdf 2014.

[86] Jia, H.; Wang, P. Enzymatic synthesis of galactosyl lactic ethyl ester and its polymer for use as biomaterials. *J. Biotechnol.* 2007, 132, 314–317.

[87] Reuter, S.; Rusborg Nygaard, A.; Zimmermann, W. ß-Galactooligosaccharide synthesis with ß-galactosidases from *Sulfolobus solfataricus*, *Aspergillus oryzae*, and *Escherichia coli*. *Enzyme Microb. Technol.* 1999, 25, 509–516.

[88] Pérez-Sánchez, M.; Cortés Cabrera, Á.; García-Martín, H.; Sinisterra, J.V.; García, J.I.; Hernáiz, M.J. Improved synthesis of disaccharides with *Escherichia coli* ß-galactosidase using bio-solvents derived from glycerol. *Tetrahedron* 2011, 67, 7708–7712.

[89] Irazoqui, G.; Giacomini, C.; Batista-Viera, F.; Brena, B.M.; Cardelle-Cobas, A.; Corzo, N.; Jimeno, M.L. Characterization of galactosyl derivatives obtained by transgalactosylation of lactose and different polyols using immobilized ß-galactosidase from *Aspergillus oryzae*. *J. Agric. Food Chem.* 2009, 57, 11302–11307.

[90] Hernaiz, M.J.; Crout, D.H.G. A highly selective synthesis of N-acetyllactosamine catalyzed by immobilised ß-galactosidase from *Bacillus circulans*. *J. Mol. Catal. - B Enzym.* 2000, 10, 403–408.

[91] Farkas, E.; Thiem, J. Enzymatic synthesis of galactose-containing disaccharides employing ß-galactosidase from *Bacillus circulans*. *European J. Org. Chem.* 1999, 1999, 3073–3077.

[92] Park, Y.K.; De santi, M.S.S.; Pastore, G.M. Production and characterization of ß-galactosidase from *Aspergillus oryzae*. *J. Food Sci.* 1979, 44, 100–103.

[93] Dragosits, M.; Pflügl, S.; Kurz, S.; Razzazi-Fazeli, E.; Wilson, I.B.H.; Rendic, D. Recombinant Aspergillus ß-galactosidases as a robust glycomic and biothecnological tool. *Appl. Microbiol. Biotechnol.* 2014, 98, 3553–3567.

[94] Kim, C.S.; Ji, E.S.; Oh, D.K. A new kinetic model of recombinant ß-galactosidase from *Kluyveromyces lactis* for both hydrolysis and transgalactosylation reactions. *Biochem. Biophys. Res. Commun.* 2004, 316, 738–743.

[95] Cornish Bowden, A. *Reactions of more than one substrate. In Fundamentals of Enzyme Kinetics*; Athel Cornish-Bowden, Ed.; Portland Press: London, 1995; pp. 129–158 ISBN 1 85578 072 0.
[96] Abada, E.A. Application of Microbial Enzymes in the Dairy Industry; Elsevier Inc., 2019; ISBN 9780128132807.
[97] Domingues, L.; Lima, N.; Teixeira, J.A. Aspergillus niger ß-galactosidase production by yeast in a continuous high cell density reactor. *Process Biochem.* 2005, 40, 1151–1154.
[98] Mawson, A.J. Bioconversions for whey utilization and waste abatement. *Boiresource Technol.* 1994, 47, 195–203.
[99] Jelen, P. Dried whey, whey proteins, lactose and lactose derivative products. In *Dairy Powders and Concentrated Products*; Tamime, A.Y., Ed.; Wiley-Blackwell: Oxford, UK, 2009; pp. 255–267.
[100] Parashar, A.; Jin, Y.; Mason, B.; Chae, M.; Bressler, D.C. Incorporation of whey permeate, a dairy effluent, in ethanol fermentation to provide a zero waste solution for the dairy industry. *J. Dairy Sci.* 2016, 99, 1859–1867.
[101] Kokkiligadda, A.; Beniwal, A.; Saini, P.; Vij, S. Utilization of cheese whey using synergistic immobilization of ß-galactosidase and *Saccharomyces cerevisiae* cells in dual matrices. *Appl. Biochem. Biotechnol.* 2016, 179, 1469–1484.
[102] Sen, P.; Nath, A.; Bhattacharjee, C.; Chowdhury, R.; Bhattacharya, P. Process engineering studies of free and micro-encapsulated ß-galactosidase in batch and packed bed bioreactors for production of galactooligosaccharides. *Biochem. Eng. J.* 2014, 90, 59–72.
[103] Van De Velde, F.; Lourenço, N.D.; Pinheiro, H.M.; Bakker, M. Carrageenan: a food-grade and biocompatible support for immobilisation techniques. *Adv. Synth. Catal.* 2002, 344, 815–835.
[104] Homaei, A.A.; Sariri, R.; Vianello, F.; Stevanato, R. Enzyme immobilization: An update. *J. Chem. Biol.* 2013, 6, 185–205.
[105] Cardelle-Cobas, A.; Olano, A.; Irazoqui, G.; Giacomini, C.; Batista-Viera, F.; Corzo, N.; Corzo-Martínez, M. Synthesis of oligosaccharides derived from lactulose (OsLu) using soluble and

immobilized Aspergillus oryzae ß-galactosidase. *Front. Bioeng. Biotechnol.* 2016, 4.

[106] Kosseva, M.R.; Panesar, P.S.; Kaur, G.; Kennedy, J.F. Use of immobilised biocatalysts in the processing of cheese whey. *Int. J. Biol. Macromol.* 2009, 45, 437–447.

[107] Ko, C.Y.; Liu, J.M.; Chen, K.I.; Hsieh, C.W.; Chu, Y.L.; Cheng, K.C. Lactose-free milk preparation by immobilized lactase in glass microsphere bed reactor. *Food Biophys.* 2018, 13, 353–361.

[108] Xavier, J.R.; Ramana, K.V.; Sharma, R.K. ß-galactosidase: Biotechnological applications in food processing. *J. Food Biochem.* 2018, 42, e12564.

[109] Neri, D.F.M.; Balcão, V.M.; Costa, R.S.; Rocha, I.C.A.P.; Ferreira, E.M.F.C.; Torres, D.P.M.; Rodrigues, L.R.M.; Carvalho, L.B.; Teixeira, J.A. Galacto-oligosaccharides production during lactose hydrolysis by free *Aspergillus oryzae* ß-galactosidase and immobilized on magnetic polysiloxane-polyvinyl alcohol. *Food Chem.* 2009, 115, 92–99.

[110] Gänzle, M.G. Enzymatic synthesis of galacto-oligosaccharides and other lactose derivatives (hetero-oligosaccharides) from lactose. *Int. Dairy J.* 2012, 22, 116–122.

[111] Guerrero, C.; Vera, C.; Conejeros, R.; Illanes, A. Transgalactosylation and hydrolytic activities of commercial preparations of ß-galactosidase for the synthesis of prebiotic carbohydrates. *Enzyme Microb. Technol.* 2015, 70, 9–17.

[112] Padilla, B.; Ruiz-Matute, A.I.; Belloch, C.; Cardelle-Cobas, A.; Corzo, N.; Manzanares, P. Evaluation of oligosaccharide synthesis from lactose and lactulose using ß-galactosidases from *Kluyveromyces* isolated from artisanal cheeses. *J. Agric. Food Chem.* 2012, 60, 5134–5141.

[113] Guerrero, C.; Vera, C.; Illanes, A. Optimisation of synthesis of oligosaccharides derived from lactulose (fructosyl-galacto-oligosaccharides) with ß-galactosidases of different origin. *Food Chem.* 2013, 138, 2225–2232.

[114] Martínez-Villaluenga, C.; Cardelle-Cobas, A.; Olano, A.; Corzo, N.; Villamiel, M.; Jimeno, M.L. Enzymatic synthesis and identification of two trisaccharides produced from lactulose by transgalactosylation. *J. Agric. Food Chem.* 2008, 56, 557–563.

[115] Guerrero, C.; Vera, C.; Plou, F.; Illanes, A. Influence of reaction conditions on the selectivity of the synthesis of lactulose with microbial ß-galactosidases. *J. Mol. Catal. B Enzym.* 2011, 72, 206–212.

[116] Shen, Q.; Yang, R.; Hua, X.; Ye, F.; Wang, H.; Zhao, W.; Wang, K. Enzymatic synthesis and identification of oligosaccharides obtained by transgalactosylation of lactose in the presence of fructose using ß-galactosidase from *Kluyveromyces lactis*. *Food Chem.* 2012, 135, 1547–1554.

[117] Vera, C.; Guerrero, C.; Wilson, L.; Illanes, A. Synthesis of propyl-ß-D-galactoside with free and immobilized ß-galactosidase from *Aspergillus oryzae*. *Process Biochem.* 2017, 53, 162–171.

[118] Vera, C.; Guerrero, C.; Wilson, L.; Illanes, A. Synthesis of butyl-ß-D-galactoside with commercial ß-galactosidases. *Food Bioprod. Process.* 2017, 103, 66–75.

[119] Ismail, A.; Ghoul, M. Enzymatic synthesis of butylglycosides by glycosidases. *Biotechnol. Lett.* 1996, 18, 1199–1204.

[120] Beecher, J.E.; Andrews, A.T.; Vulfson, E.N. Glycosidases in organic solvents: II. Transgalactosylation catalysed by polyethylene glycol-modified ß-galactosidase. *Enzyme Microb. Technol.* 1990, 12, 955–959.

[121] Vera, C.; Guerrero, C.; Wilson, L.; Illanes, A. Optimization of reaction conditions and the donor substrate in the synthesis of hexyl-ß-D-galactoside. *Process Biochem.* 2017, 58, 128–136.

[122] Mladenoska, I.; Winkelhausen, E.; Kuzmanova, S. Transgalactosylation/hydrolysis ratios of various? ß-galactosidases catalyzing alkyl- ß-galactoside synthesis in single-phased alcohol media. *Food Technol. Biotechnol.* 2008, 46, 311–316.

[123] Vetere, A.; Medeot, M.; Campa, C.; Donati, I.; Gamini, A.; Paoletti, S. High-yield enzymatic synthesis of O-allyl ß-d-galactopyranoside. *J. Mol. Catal. B Enzym.* 2003, 21, 153–156.

[124] Hart, J.B.; Kröger, L.; Falshaw, A.; Falshaw, R.; Farkas, E.; Thiem, J.; Win, A.L. Enzyme-catalysed synthesis of galactosylated 1D- and 1L-chiro-inositol, 1D-pinitol, myo-inositol and selected derivatives using the ß-galactosidase from the thermophile Thermoanaerobacter sp. strain TP6-B1. *Carbohydr. Res.* 2004, 339, 1857–1871.

[125] Lee, S.E.; Jo, T.M.; Lee, H.Y.; Lee, J.; Jung, K.H. ß-Galactosidase-catalyzed synthesis of galactosyl chlorphenesin and its characterization. *Appl. Biochem. Biotechnol.* 2013, 171, 1299–1312.

[126] Stevenson, D.E.; Furneaux, R.H. High-yield syntheses of ethyl and using Streptococcus thermophilus. *Enzyme Microb. Technol.* 1996, 18, 513–518.

[127] Binder, W.H.; Kählig, H.; Schmid, W. Galactosylation by use of ß-galactosidase: Enzymatic syntheses of disaccharide nucleosides. *Tetrahedron: Asymmetry* 1995, 6, 1703–1710.

[128] Zervosen, A.; Nieder, V.; Gutiérrez Gallego, R.; Kamerling, J.P.; Vliegenthart, J.F.G.; Elling, L. Synthesis of nucleotide-activated oligosaccharides by ß-galactosidase from Bacillus circulans. *Biol. Chem.* 2001, 382, 299–311.

[129] Scheckermann, C.; Wagner, F.; Fischer, L. Galactosylation of antibiotics using the ß-galactosidase from *Aspergillus oryzae*. *Enzyme Microb. Technol.* 1997, 20, 629–634.

[130] Aragón, J.J.; Cañada, F.J.; Fernández-Mayoralas, A.; López, R.; Martín-Lomas, M.; Villanueva, D. A direct enzymatic synthesis of ß-D-galactopyranosyl-D-xylopyranosides and their use to evaluate rat intestinal lactase activity *in vivo*. *Carbohydr. Res.* 1996, 290, 209–216.

[131] Carevic, M.; Velickovic, D.; Stojanovic, M.; Milosavic, N.; Rogniaux, H.; Ropartz, D.; Bezbradica, D. Insight in the regioselective enzymatic transgalactosylation of salicin catalyzed by ß-galactosidase from *Aspergillus oryzae*. *Process Biochem.* 2015, 50, 782–788.

[132] Zeleny, R.; Altmann, F.; Praznik, W. A capillary electrophoretic study on the specificity of ?-galactosidases from *Aspergillus oryzae, Escherichia coli, Streptococcus pneumoniae,* and *Canavalia ensiformis* (Jack Bean). *Analyt. Biochem.* 1997, 246, 96–101.

[133] Rodríguez, E.; Francia, K.; Brossard, N.; García Vallejo, J.J.; Kalay, H.; van Kooyk, Y.; Freire, T.; Giacomini, C. Immobilization of ß-galactosidase and a-mannosidase onto magnetic nanoparticles: A strategy for increasing the potentiality of valuable glycomic tools for glycosylation analysis and biological role determination of glycoconjugates. *Enzyme Microb. Technol.* 2018, 117, 45–55.

[134] Iskratsch, T.; Braun, A.; Paschinger, K.; Wilson, I.B.H. Specificity analysis of lectins and antibodies using remodeled glycoproteins. *Anal. Biochem.* 2009, 386, 133–146.

In: Beta-Galactosidase
Editor: Eloy Kras

ISBN: 978-1-53615-605-8
© 2019 Nova Science Publishers, Inc.

Chapter 3

GALACTO-OLIGOSACCHARIDE SYNTHESIS BY TRANSGALACTOSYLATION ACTIVITY OF β-GALACTOSIDASE: RECENT TRENDS, CHALLENGES AND FUTURE PERSPECTIVES

Milica Simović[1,*], *Marija Ćorović*[1]
and Dejan Bezbradica[1], *Ana Milivojević*[2]
and Katarina Banjanac[2]

[1]Department of Biochemical Engineering and Biotechnology, Faculty of Technology and Metallurgy, University of Belgrade, Belgrade, Serbia

[2]Innovation Center, Faculty of Technology and Metallurgy, University of Belgrade, Belgrade, Serbia

---

[*] Corresponding Author's E-mail: mcarevic@tmf.bg.ac.rs.

Abstract

The growing interest in functional foods and their beneficial effects on human health, yielded due interest in the field of obtaining novel bioactive ingredients featuring enhanced physiological and physical/chemical characteristics. Accordingly, the considerable attention of the scientific community during the last decades was attributed to β-galactosidase synthetic activity. Namely, under the specific conditions, β-galactosidase catalyzes reaction of transgalactosylation resulting in the formation of a diverse mixture of highly valuable products named galacto-oligosaccharides (GOS). Owing to the specific structure, GOS are primarily recognized as products with the pronounced prebiotic activity and favorable impact on overall human well-being, however, their wide application in the food and pharmaceutical industry is additionally ensured by their high thermal and acid stability, excellent taste quality, low sweetness and caloric value.

Throughout this chapter, different aspects related to the recent advances in GOS production and purification, as well as their application with emphasis on their prebiotic role, will be covered. Likewise, different challenges and future perspectives will be discussed.

Keywords: β-galactosidase, transgalactosylation, galacto-oligosaccharides, GOS

Introduction

Galacto-oligosaccharides (GOS) represent a complex mixture of non-digestible carbohydrates, most commonly composed of 2-8 monosaccharide units (Figure 1), optionally terminal glucose and several remaining galactose units [1]. These compounds are markedly different in their composition, type of bonds and degree of polymerization, which consequently can cause differences in their physical, chemical and functional properties [2-3]. However, they are commonly recognized as a highly valuable diverse group of prebiotic compounds, mostly due to their ability to selectively stimulate the proliferation and activity of beneficial intestinal microbiota (bifidobacteria and lactobacilli) [1, 4]. Additionally, there are several other benefits to the overall health such as: the improvement of lactose digestion and mineral absorption, reduction of serum cholesterol level, inhibition of intestinal pathogen growth, reducing cancer risk and enhancement of a host's immune system; all of which can be associated with the consumption of GOS [5-6]. Moreover, their excellent thermal and acid stability, low sweetness and caloric value, and their notable contribution to the taste quality of products, make GOS an ideal functional ingredient of infant milk formulas, dairy products, beverages and lately animal feed [7-8].

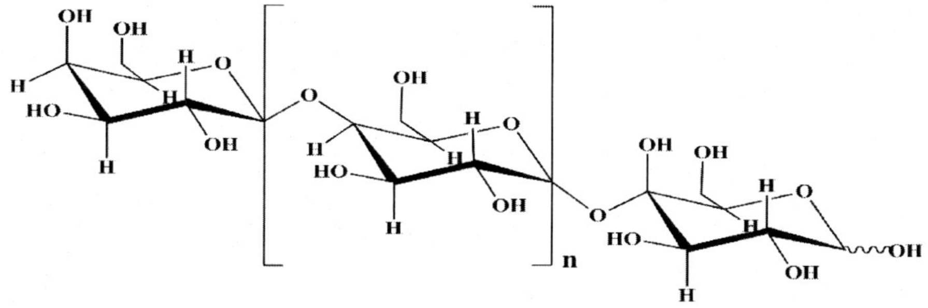

Figure 1. Schematic presentation of galacto-oligosaccharides.

When detected for the first time in 1950s during the lactose hydrolysis process, GOS were considered an undesirable by-product, which was found to have an adverse effect to the quality of the final product due to the low levels of sweetness [9]. However, the similarity with human milk oligosaccharides (HMO) and the ability to mimic their functions and properties have renewed the interest in these compounds [1-2]. During the past few decades numerous GOS synthesis and purification methods have been examined [6, 10-12], and they are currently commercially produced worldwide. The industrial production of GOS began in Japan in 1970s, where they were used predominantly infant nutrition [7]. However, since their exceptional functional and physical/chemical characteristics have been proven repeatedly, their production has been increasing rapidly all over Europe and America, reaching around 175 kt nowadays with significantly broader fields of application (functional food and feed, as well as cosmetics) [12].

ENZYMATIC SYNTHESIS OF GALACTO-OLIGOSACCHARIDES

Galacto-oligosaccharides are present in nature as a constituent of mammalian milks [13]. However, nowadays they are most commonly obtained through the process of enzymatic synthesis. It is also notable that there are examples of GOS chemical synthesis which involve the use of inorganic acids in harsh reaction conditions [14-15], as well as examples of the synthesis catalyzed by glycosyltransferases, but these methods proved to be generally unfavourable for various reasons [16-17].

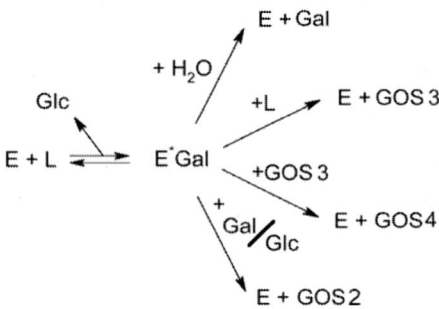

Figure 2. Reactions within production of galacto-oligosaccharides: E- enzyme, L – lactose, Glc - glucose, Gal – galactose, EGal* - enzym-galactosyl complex, GOS2 – disaccharides (other than lactose), GOS (trisaccharides), GOS4 (tetrasaccharides).

Even though the glycosyltransferases are genuinely efficient and show high stereo- and regio-selectivity, they require specific substrates (e.g., sugar nucleotides) to be effective which, along with their high price, limits their broader use [18]. β-Galactosidases, on the other hand, are poorly selective, primarily hydrolytic enzymes, but represent relatively cheap and readily available enzymes that do not require any additional steps for substrate preparation, and use a widely available substrate – lactose (pure or as a part of complex dairy substrates) [2, 19-20].

GOS synthesis catalyzed by β-galactosidase is a quite complex process, since β-galactosidase catalyses two simultaneous reactions - lactose hydrolysis and transgalactosylation. Therefore, multiple parallel steps take place during the course of reaction [2], which result in the formation of several products: different GOS molecules, glucose and galactose (Figure 2). In terms of achieving high GOS yields, good understanding of reaction mechanism and thorough optimization of process parameters are required.

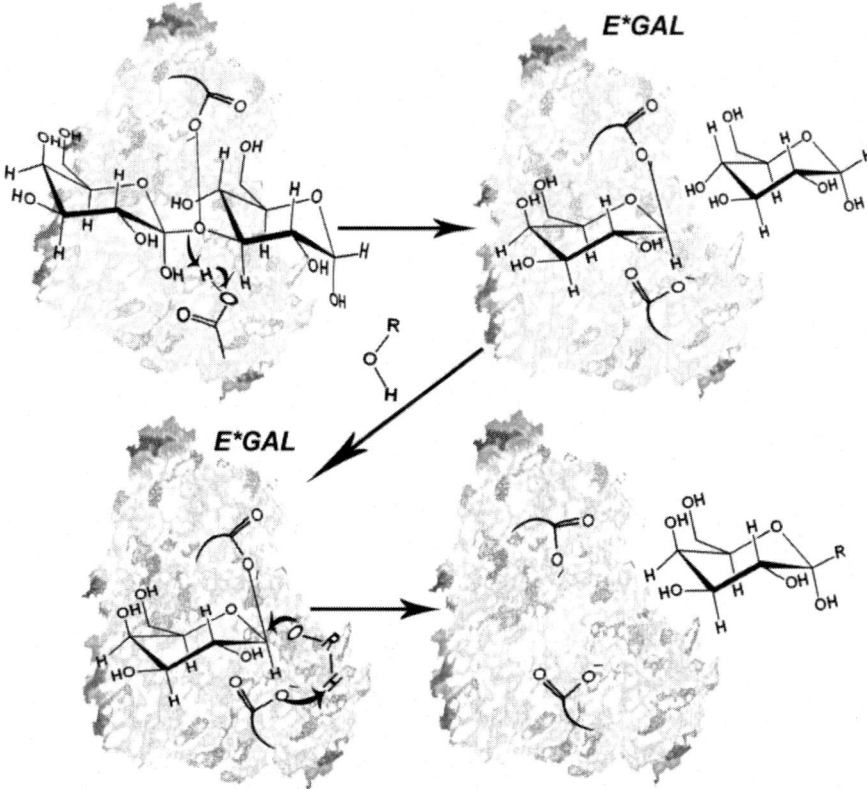

Figure 3. β-Galactosidase reaction mechanism. ROH in the second phase could be any molecule containing free OH group (water, lactose, galactose, GOS)[2].

When observing the reaction mechanism, it can be noted that the first stage of both reactions (hydrolysis and transgalactosylation) is common. During this stage, lactose molecules approach the active site of the enzyme, where different amino acid residues participate in the binding and/or conversion of the substrate both directly and indirectly [21]. Namely, in case of β-galactosidases two glutamic acid residues exhibit the main roles in the reaction mechanism being the proton donor and nucleophile/base [22-24], while several other amino acid residues support the reaction by forming hydrogen bonds and placing sugar moieties in their proper positions [23-25].

The process of lactose entering the active site is succeeded the covalent bond formation between the galactosyl moiety of lactose and the nucleophile of the enzyme (the glutamic acid residue), after which the enzyme-galactosyl (EGal*) complex is formed [24]. Simultaneously, the second glutamic acid residue transfers a proton onto the glucose moiety of lactose thus releasing it from the enzyme active site. In the second stage, a nucleophilic attack of the acceptor molecule onto the enzyme-galactosyl complex occurs, whereby, depending on the present acceptors, the reaction course diverges into two directions: transgalactosylation or hydrolysis. In case a galactose moiety acceptor is a molecule of water, the reaction of hydrolysis occurs and the galactose molecule is released. Otherwise, if a galactose acceptor is another compound with a free hydroxyl group from the reaction mixture (e.g., other sugar), a transgalactosylation reaction will follow [2]. Accordingly, all sugars present in the reaction mixture can act as galactosyl moiety acceptors owing to the free hydroxyl group presence. Hence, the final reaction mixture may contain a wide range of possible products: di- (GOS2), tri- (GOS3), tetra- (GOS4) and higher oligosaccharides with β-glycosidic links between the monosaccharide units. However, the composition of obtained products is highly dependent on the characteristics and origin of the employed enzyme, and the reaction conditions during the enzymatic synthesis [10, 26].

It must be noted that all reactions within the proposed mechanism act as reversible reactions, and consequently, the obtained GOS could be subjected to the hydrolysis process, too. Thus, GOS can be regarded as a mid-product rather than final reaction product. This fact indicates that the above-mentioned reaction is highly kinetically-controlled [27], and after a long enough period of time, only the final reaction products (glucose and galactose) will be obtained. Therefore, the course of the reaction should be thoroughly examined, and special attention should be paid to the reaction time optimization (Figure 4). During the first stage of the reaction, while lactose concentration is still high, the transgalactosylation reaction apparently prevails [27-28]. There is a noticeable decrease in lactose concentration, which is primarily spent on the GOS synthesis, and only in a small part on hydrolysis. As the reaction proceeds, the concentration of lactose is becoming lower and less available as a galactosyl acceptor, and the reaction of GOS synthesis stagnates. On the other hand, a steady increase in the concentration of hydrolysis products (glucose and galactose) can be noticed [27-28]. If the reaction lasts long enough the complete hydrolysis of lactose and GOS will occur. Accordingly, it is extremely important to stop the reaction at the optimum reaction time in order to obtain the maximum product yield. Bearing this in mind, the maximum yields of the transgalactosylation reaction are not particularly high and vary in the range of 20-40% GOS [11], with occasional exceptions. Nevertheless, in addition to the decisive effect on product yield, it has been observed that the reaction time also affects the composition of the obtained GOS mixture - in terms of both polymerization degree and type of bonds within molecules [28]. Naturally, synthesis of shorter GOS prevails in the first reaction stages since they represent the required substrates for higher GOS, and during the reaction course their intense growth is slowly reduced during their conversion into higher GOS, whose production becomes dominant appropriately. Finally, when the reaction of hydrolysis becomes more significant, overall GOS concentration decreases following the opposite pattern.

Although numerous examples of GOS production have been described in scientific publications so far [12, 26, 29], a large number of factors simultaneously influencing the yield and the galacto-oligosaccharides structure prevent the definition of the general optimal conditions for the GOS synthesis. Some potential strategies for obtaining high yields of desired GOS include: selecting an adequate enzyme preparation that exhibits a high transgalactosylation activity and ensures the reduction of substrate inhibition [26], using highly concentrated lactose solutions or systems with reduced water activity [30-31], working at higher temperatures in terms of achieving better lactose solubility [32-33], performing the reaction in the most suitable reactor configurations in terms of achieving high transgalactosylation performance while decreasing the substrate inhibition and optimum time control [34-36]. In the next sections we will attempt to describe the effects of the most relevant parameters on process productivity.

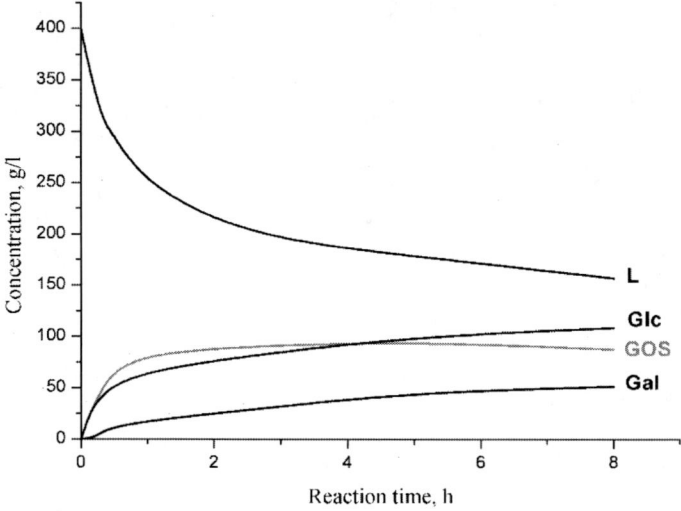

Figure 4. Characteristic concentration profiles of reaction mixture components during the reaction course. L - lactose, Glc - glucose, Gal - galactose, GOS - total galacto-oligosaccharides

ENZYME SOURCE

As previously mentioned, the primary biological function of β-galactosidase is the lactose hydrolysis reaction [37]. Therefore, the ability of an enzyme to perform the reaction of transgalactosylation depends to the considerable extent on the origin of the employed enzyme [38]. It is believed that the enzyme structure plays a key role in substrate specificity and its ability to synthesize GOS [39]. These claims are somewhat confirmed in several experiments featuring genetically-modified enzymes, which showed alleviated transgalactosylation potential when certain sequences were substituted or deleted [40-41]. Furthermore, it is known that some enzymes display less affinity towards GOS degradation and glucose and galactose inhibition, thus enable production of higher GOS yields [12].

In addition to the above mentioned, the source of the enzyme also exhibits a significant influence on the structure of the obtained GOS. Namely, certain β-galactosidases have a higher affinity towards the formation of certain types of glycosidic bonds, thus forming GOS mixtures with different β-glycosidic bonds [39]. Also, the degree of GOS polymerization is highly dependent on the nature of the enzyme [39]. Up to the date, GOS synthesis has been tried with various enzymes under different conditions, which is described in great detail in various review papers [2, 6, 10-11, 26, 29, 42]. However, among the most exploited β-galactosidases both on laboratory and industry scale, owing to their commercial availability and recognized GRAS (generally regarded as safe) status, are enzymes originated from *Aspergillus oryzae*, *Bacillus circulans* and *Klyveromyces lactis*. These highly available enzymes are good examples to illustrate the enzyme specificity towards GOS synthesis as a function of enzyme origin. Namely, β-galactosidase from *B. circulans* produces predominantly β-(1→4) linked GOS with a polymerization degree 2-5, and the approximate total GOS yields are around 40% [1]. At the same time, *K. lactis* β-galactosidase produces smaller amounts (approximately 30%) of mainly β-(1→6) linked GOS with polymerization degree 2-4, with dominant disaccharide fraction [12]. On the other hand, β-galactosidase from *A. oryzae* does not produce disaccharides, but almost exclusively β-(1→4), β-(1→6), and β-(1→3) linked tri-saccharides (GOS3) and tetra-saccharides (GOS4). The highest achieved yields using this enzyme have not exceeded 30% [28, 43]. To the present, the highest achieved galacto-oligosaccharides concentration (315 g/l) was obtained by conversion of concentrated lactose solution (600 g/l) using thermostable β-galactosidase from *Sulfolobus solfataricus* [44]. Therefore, owing to their possibility of carrying out the reaction at higher temperatures and enabling reactions with highly concentrated lactose solutions, a new kind of β-galactosidases from thermophilic microorganisms has come under the spotlight. Moreover, special attention is drawn to the prospective use of probiotic sources (lactic acid bacteria) β-galactosidases in view of achieving targeted GOS production, thus providing GOS with potentially pronounced physiological activity [45-53]. The great example of this is the commercial production of GOS by virtue of β-galactosidase from *Bifidobacterium bifidum* (Bimuno, Calsado Ltd.)[7].

Interestingly, the ability of certain soluble enzyme to synthesize galacto-oligosaccharides, however, exists in conjunction with other reaction factors, and is susceptible to the change under altered conditions (e.g., enzyme immobilization). Hence, different strategies for β-galactosidase immobilization (physical absorption, covalent immobilization, entrapment and enzyme cross-linking) were examined in terms of achieving higher productivity, process cost-effectiveness, enzyme stability and selectivity improvement, and suppression of potential substrate (glucose and galactose) inhibition [36, 54-62]. Even though several papers thoroughly review achievements of various immobilized preparations in transgalactosylation reactions [10-11], none of these immobilized preparations have found their application in any of the existing commercial processes yet. Finally, after the comprehensive insight in synthesis by isolated enzymes, both in soluble or immobilized form, it must be mentioned that the GOS synthesis is conducted with considerable success using microbial whole cells (living, resting, permeabilized and also immobilized cells)[17, 63-64]. More importantly this method is commercially exploited, namely, Nissin Sugar Co., Ltd. is currently producing GOS using permeabilized cells of *Cryptococcus laurentii* (Cup-Oligo in powder and syrup form) [7].

EFFECT OF LACTOSE CONCENTRATION

Looking at the very mechanism of the transgalactosylation reaction (Figure 3), it can be noticed that the initial lactose concentration is a key factor influencing the process of GOS synthesis. The influence of lactose concentration on GOS production can be analysed from two main aspects: acceptor abundance and lowering water activity. Namely, with the increase of lactose concentration, the increase of potential acceptor of galactose molecules occurs, and GOS production is promoted at the expense of hydrolysis. On the other hand, high levels of lactose reduce water activity [65], which simultaneously reduces the reaction rates of lactose and just synthesized GOS hydrolysis [30-31]. Besides the total GOS yield, the increase of lactose concentration affects the polymerization degree of obtained GOS, ensuring larger portions of higher GOS in the final product mixture [10]. Also, there are some indications that the increase of lactose concentration might reduce product inhibition (usually competitive galactose inhibition), and is favourable to the enzyme from the stability point of view [66]. It should be noted that, regardless of the enzyme origin and other reaction parameters, scientific data on the effect of lactose concentration are consistent, stating that the concentrations of lactose above 30% favour the reaction of transgalactosylation [12, 29]. The main obstacle for achieving higher initial lactose concentrations is rather poor lactose solubility (189 g/l at 25ºC, 251 g/l at 40ºC, and 372 g/l at 60ºC). Therefore, lactose is usually dissolved at higher temperatures to achieve supersaturate lactose solutions (with less than 2.1 super-saturation factor) that are stable upon cooling to the optimum reaction temperature, and subjected to the reaction [12, 29]. However, maximum GOS yields seem to increase proportionally until the lactose concentration reaches 30%, and afterwards the increase in GOS yields becomes less pronounced [1].

EFFECT OF TEMPERATURE AND PH

The effect of temperature on the GOS synthesis is considered to be important as well, although the scientific publications on this topic show certain inconsistencies. Generally, the increase in temperature is regarded as a positive effect considering its influence on reaction rates (hydrolysis and transgalactosylation) and increased lactose solubility, which enables higher initial lactose concentrations [38]. Several studies have shown that the increase in temperature can favour the transgalactosylation reaction, since the effect of temperature on the rate of the synthesis reaction is more pronounced than on the rate of hydrolysis reaction [67]. On the other hand, increased temperatures have an adverse effect on enzyme activity [68]. Hence, the utilization of thermophilic β-galactosidases proves to be an attractive choice [26, 44], although these sources are not commercially available for industrial applications and are not approved in food and pharmaceutical applications. Iwasaki and associates concluded that the effect of temperature is an extremely important factor when we compare the transgalactosylation activity of enzymes of different origin, but that for individual enzymes the temperature for achieving maximum yields is quite wide, and it usually coincides with the optimal temperature for the reaction of hydrolysis [69]. A similar situation is with the pH value of the reaction mixture. The effect of the pH on the transgalactosylation reaction can be significant and depends primarily on the nature of the enzyme. There are indications that a pH change may cause the increase in the enzyme selectivity towards the transgalactosylation reaction, however the general conclusion could not be reached.

ALTERNATIVE WAYS TO ACHIEVE HIGHER GOS YIELDS

The optimization of the conditions may increase the GOS yield to some extent, but the use of available β-galactosidases still provide relatively modest yields. In order to overcome this problem, and considering the genuine progress of biotechnology, an increase of the enzyme affinity towards the transgalactosylation reaction has been proposed on the basis of genetic and protein engineering. For example, Jorgensen and associates were able to increase affinity for the transgalactosylation reaction 10 times and reduce the galactose inhibitory activity significantly at the same time by virtue of shortening the N-terminal end of β-galactosidase originated from B. bifidum [41]. Similar results are achieved by targeted mutations (amino acid substitution in the protein chain) on β-galactosidase from Geobacillus stearotermophilus [70]. More importantly, the increased knowledge on the behaviour of enzymes will enable controlled genetic engineering influence not only on product yields, but also on the structure and function of the obtained GOS.

Since the availability of lactose and the reduction of water activity are key parameters used to optimize the GOS synthesis process, the use of highly concentrated solutions represents a safe way of accomplishing high GOS yields. In order to obtain highly concentrated lactose solutions, the reactions should be carried out at very high temperatures which require the detection and isolation of novel thermostable β-galactosidases that will be highly efficient for GOS synthesis in such conditions [26, 42]. However, this approach is quite challenging, as any potential thermostable β-galactosidases producer might turn out to be unsafe for application in food industry. Nevertheless, genetic engineering might help to resolve this, and targeted thermophilic strain genes may be cloned to be safe and earn their GRAS status [71].

Reduction of water activity, besides using highly concentrated lactose solutions, may be provided by replacing water or at least a portion of the aqueous phase with organic solvents [72-73] or by adding various salts that will not adversely affect enzyme stability [74]. However, despite some promising results, the restrictions on the use of organic solvents, poor lactose solubility and instability of enzymes in the presence of solvents and certain ions, showed that further optimizations should be performed. Certain ionic liquids could be used as an adequate substitution for organic solvents [75].

During the reaction, the GOS synthesis inhibition may occur as a result of the formation of certain reaction products (glucose and/or galactose) [33, 43]. Therefore, it is considered that the inhibition can be prevented by removing such products, while simultaneously achieving a significant improvement in the process yields, since somewhat purified products are obtained. There are various strategies which can be used to remove monosaccharides from the mixture. The continuous removal of the formed glucose by virtue of different enzymes (e.g., glucose oxidase and catalase) or whole cells was thoroughly examined. One common example of the removal of glucose entails the use of permeabilized cells as a way of achieving simultaneous GOS synthesis and selective monosaccharide fermentation [76-77]. Nowadays, permeable cells are used more often instead of expensive commercial and purified enzymatic preparations, since their use reduces the costs of isolating and purifying the enzymes significantly. Furthermore, they can also be used repeatedly, which reduces the cost of the entire GOS synthesis process [77]. These cells can retain their viability, and as such can simultaneously perform the transgalactosylation process and the process of glucose fermentation, which leads to an increase in the yield of the reaction. On the other hand, monosaccharides can continually be removed by initiating the reaction in membrane reactors carefully chosen to retain enzyme and oligosaccharides [78-79]. In all these reactions, disaccharides (GOS2) yield is significantly reduced due to the lack of monosaccharides which would act as acceptors. However, the tetrasaccharide (GOS4) concentration was significantly higher than in the control reaction, while the trisaccharide (GOS3) yield remained unchanged. Nonetheless, the achieved yields of GOS synthesis still remain modest, and further mixture purification is often advised.

PURIFICATION OF GALACTO-OLIGOSACCHARIDES

As previously mentioned, the main problem of GOS production are rather low reaction yields, usually in the range 20-40% (with rare exceptions). Despite the thorough reaction optimization process, the concentrations of residual lactose and hydrolysis products (glucose and galactose) still remain unacceptably high. Therefore, further purification of GOS from the obtained reaction mixture is necessary for the successful application. For example, lactose removal is important in terms of avoiding crystallization, and ensuring that the products can find broader use, while also reducing the possibility of lactose maldigestion and intolerance [80] Additionally, the removal of monosaccharides leads to lower sugar content and caloric value of obtained products, which can be suitable for high share of population suffering from obesity and diabetes [5].

However, due to the desired structural complexity of the product and noticeable similarities in carbohydrate structures within the reaction mixture, GOS purification proves to be a rather difficult task. Hence, in order to achieve the required purity by fractionation of saccharide molecules, several advanced purification techniques (membrane separation, chromatography, solvent fractionation and bioconversion) have been examined (Figure 5).

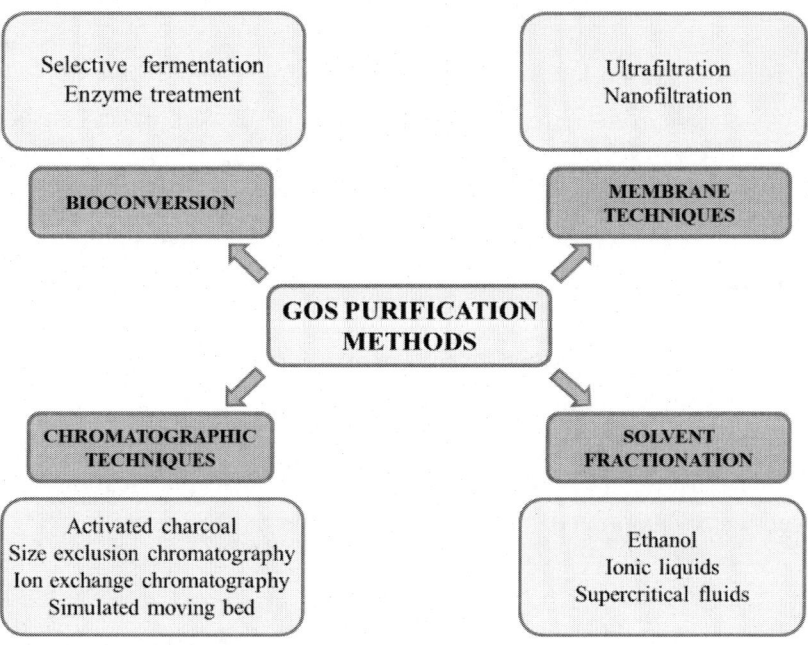

Figure 5. GOS purification methods.

FRACTIONATION USING SOLVENTS

Solubility data for saccharides in different solvents is well documented. Considering the differences in solubility, selective precipitation of saccharides was proposed as a compelling strategy for GOS purification. Although this simple and cost-effective fractionation technique is successfully utilized for lactose precipitation and purification [81], examples of GOS purification using the selective precipitation are scarce and include the usage of ethanol [82]. Namely, purification by fractional precipitation with 90% ethanol in two sequential cycles yielded an increase in GOS content from 15% to 75%, while simultaneously the reduction of monosaccharides dropped from 48% to 4%. However, despite the good results in terms of purity, recovery of GOS was quite poor (below 10%). In order to increase recovery yields, the initial concentrations of saccharide mixture should be increased, which affects the purificaiton process negatively because of the non-selective precipitation [82]. Therefore, an optimal balance between two outputs should be established. Also, unnecessary dilutions that have a negative effect on process costs and complexity should be avoided. In conclusion, it should be used in combination with other fractionation techniques despite all the advantages of solvent precipitation process (simplicity, low cost, and scalability).

A different approach to solvent fractionation includes the use of green solvent alternatives, such as ionic liquids, supercritical fluids and pressurized fluids, which tend to improve the selectivity and efficiency of the process while simultaneously lowering the consumption of volatile solvents. Even though these versatile solvents are nowadays widely applied for similar purposes [83-84], their application in GOS purification has not yet been widely examined.

CHROMATOGRAPHIC TECHNIQUES

Activated Charcoal Adsorption

One of the simplest and cheapest procedures for GOS purification is the adsorption using activated charcoal, a method widely used in sugar refining industry. Activated charcoal is a rather cheap sorbent material, an easily regenerated large surface area that retains saccharides due to its highly non-polar/hydrophobic nature [85]. Saccharides will be adsorbed according to their molecular weight, owing to the fact that hydrophobic sugars (with higher polymerization degree) are more susceptible to the adsorption onto activated charcoal [29, 86]. However, it must be noted that the configuration of molecules plays an important role in adsorption as well, since planar molecules are generally more retained in comparison to the branched ones [87-88]. Nevertheless, the GOS are more adsorbed onto the activated charcoal than other small saccharides, which is a prerequisite for their separation. Most commonly, purification consists of three main steps: adding the activated charcoal to the saccharide mixture, washing non-retained compounds (monosaccharides) using pure water, and finally selective desorption of the adsorbed sugars using ethanol gradients. For example, during the activated charcoal purification of typical product mixture (GOS, monosaccharides and lactose), monosaccharides and lactose were removed by elution with low concentration ethanol solutions (between 1% and 15%, v/v), whilst GOS were desorbed by elution with higher concentration ethanol solutions (up to 50%, v/v) [29]. With the increase of the ethanol concentration in the solvent, the purity of the obtained GOS product increases as well, but at the expense of the obtained GOS yields [73]. In conclusion, this treatment proved to be a simple, inexpensive and rapid method, yet the main disadvantages of activated charcoal treatments are poor resolution, leading to an incomplete removal of lactose and, poor GOS yields. However, being easily regenerated and cheap, with great potential to desalt solutions, this method is used in the first phases of commercial GOS purification (e.g., Oligomate) [86].

Size-Exclusion Chromatography

Size-exclusion chromatography (SEC) is an effective purification method that allows separation of saccharides based on their size (and in some cases molecular weight). It is primarily used for semi- or preparative fractionation of GOS on laboratory level. The fractionation of saccharides is achieved by the differential exclusion or inclusion of saccharides as they pass through the stationary phase consisting of porous cross-linked polymeric gels or beads. The porous stationary phase is available in a wide range of materials: dextran (Sephadex), agarose (BioGel-A, Sepharose), and polyacrylamide (Sephacryl or BioGel-P), among others. More importantly, they can also be found in a wide range of pore sizes. Several authors have used SEC to remove GOS from the saccharide mixtures using polyacrylamide gels [86, 89] or dextran gels [90]. Using the polyacrylamide gel (BioGel-P2) which allows the separation of compounds with molecular weight in the range of 100-1800 Da, Hernández and co-workers obtained highly pure GOS fractions (degree of polymerization up to 8) with high degree of recovery (81-92%) [86]. Therefore, it was concluded that this method guarantees a saccharides fractionation with high levels of both purity and yields. Nevertheless, the main flaw of this method is significant product dilution, demanding additional steps of water (solvent) removal.

Ion Exchange Chromatography

Ion exchange chromatography (IEC) has mostly been used as a chromatographic technique for oligosaccharide fractionation on semi- and preparative level [91], while it is predominantly employed in demineralization of oligosaccharide solutions on an industrial level (e.g., Oligomate and Purimune) [29]. Ion exchange resins are usually based on poly (styrene–divinylbenzene) beads that ensure great potential for separation of carbohydrates, while at the same time show great resistance to different extern influences [92]. It is well documented that resin acociated ions (H^+, Na^+, K^+, Ag^+, Ca^{2+}, Pb^{2+}) play a substantial role in saccharide fractionation, and its proper choice can significantly improve the efficiency of separation process [92-93]. However, the mechanism of saccharides separation is not typical ion exchange, since it includes several different separation mechanisms. This means that the separation of the varying chain-length saccharides is a consequence of the size exclusion mechanism, whereas the separation of the same chain length saccharides (e.g., monosaccharides) is quite complex. It includes an ion exchange contribution, electrostatic repulsion/attraction, hydrophobic interactions, steric hindrance and van der Waals forces, among others [94]. Monovalent ions only form weak complexes with sugars, and this resin separation mechanism is mainly based on the combination of size exclusion and restricted diffusion effects, whereas ions with increasing ionic valence may form the strongest complexes with sugars [91]. However, strong cationic exchangers featuring H⁺ ions were selected as most effective in the GOS purification. Namely, commercial resins Dowex 50WX4, and Diaion UBK 530 both in H⁺ form were used for GOS purification from the commercial mixture Vivinal® GOS in two separate studies [92-93]. On the other hand, the best purification of raw GOS produced with immobilized *A. oryzae* β-galactosidase in a packed-bed reactor was achieved with Dowex 50W (Na⁺ form) [95]. Finally, it should be noted that IEC can be performed in column batch-wise and continuously in simulated moving bed (SMB) chromatography.

Simulated Moving Bed Chromatography

Simulated moving bed (SMB) chromatography is one of the most promising techniques on an industrial scale for carbohydrate fractionation, since it is able to perform very complex separations (e.g., resolution of the enantiomeric forms of a chiral compound) in a continuous regime [96-97]. It ensures increased productivity and lower consumption of eluent and energy in comparison to conventional chromatography techniques, although its design, optimization and operation costs are significantly higher [97-98]. The SMB system consists of several connected chromatographic columns featuring a complex arrangement of valves that enables the simulation of solid phase movement. The main peculiarity of this system is the counter-current flow of solid and liquid phase, without the real motion of the adsorbent [10, 29]. Instead, the movement of the adsorbent is simulated by the valves that move the position of two inlet (feed and eluent) and two outlet streams (raffinate and extract) by switching one column in the direction of the liquid phase flow at appropriate time intervals [29]. SMB chromatography is highly versatile, since it is applicable to all kinds of chromatography, although the adsorbents are usually size-exclusion, ion-exchange gels or their combination [87]. For instance, SMB system equipped with eight columns packed with a cation exchanger and a 64-port multifunctional rotary valve, yielded raffinate containing more than 99% GOS, while the extract was almost exclusively composed of mono- and disaccharides [99]. Also, fractionation of a complex mixture (lactose and human milk oligosaccharides) was successfully achieved using this method [100]. Hence, this system proved to be capable of providing both high product purity and high process yields and as such represents a good choice when GOS of high purity is required.

SELECTIVE FERMENTATION

Due to the fact that the obtained GOS reaction mixture comprises of fermentable saccharides (glucose, galactose, and to some extent lactose) and GOS that cannot be metabolized by majority of microorganisms, removal of GOS contaminants, by virtue of selective fermentation, seems like a promising solution to GOS purification problem. Therefore, the GOS purification is mainly achieved by selective fermentation using yeasts, such as: *Saccharomyces cerevisiae*, *Kluyveromyces lactis* and *K. marxianus*, since these strains have GRAS status and have been used in the food industry for decades now. *S. cerevisiae*, commonly known as brewer's/baker's yeast, is a readily available and cheap yeast strain able to efficiently ferment glucose and galactose, while lactose and GOS remain almost intact owing to the β-galactosidase absence [101-102]. In this case, glucose is completely transformed to ethanol and CO_2, while galactose consumption depends on its concentration in the saccharide mixture [101-102], as well as yeast/saccharide ratio [86, 101-102]. Taking into account the abundance of unreacted lactose, the obtained purification yields and purity are quite modest (up to 60% of GOS in mixture), and under optimal conditions the maximum 1.5 fold purification can be achieved [11]. Purity of GOS can be significantly increased if lactose is partially removed during the pre-hydrolysis step that can be performed using pronounced hydrolytic activity of β-galactosidases from *Kluyveromyces* strains [103].

On the other hand, yeasts from the *Kluyveromyces* genera are able to consume lactose and consequently eliminate all GOS contaminants. However, these enzymes are highly sensitive to osmotic stress, caused by high saccharide concentrations [11]. Therefore, in most cases, saccharide mixture has to be diluted in terms of allowing efficient consumption of contaminating saccharides and efficient purification [104]. Alternatively, high yeast concentrations are required for efficient removal of GOS contaminants. In order to achieve this, additional nutrients for intensive yeast growth (salts, minerals and yeast extract) may be required, causing additional costs for the process. To overcome these obstacles, immobilized yeast cells could be utilized to enable the necessary protection from osmotic shock, ensure its reusability and contribute to the cost-effectiveness of the process [101]. Likewise, GOS synthesis and purification could be simultaneously performed, thus alleviating the need for significant reaction mixture dilutions, since the concentrations of fermentable sugars are low in the beginning [105]. However, the main drawback of selective purification, independent of the microorganism used, is the presence of contaminants that appear during the fermentation process, which eventually may require new purification steps until the product has obtained satisfactory characteristics.

MEMRANE TECHNIQUES

Membrane technology has become one of the most important separation techniques in the past few decades [106-108]. It proves to be a convenient alternative for GOS downstream processing, since it offers numerous advantages over other previously defined techniques, such as: low energy consumption, easy integration within any existing downstream process, easy scale up, and no required additives that should be removed in later stages of purification process [109-110]. The main reason for the use of membrane separations in GOS purification process lays in the fact that it includes the retention of oligosaccharides in the concentrate and the removal of primary monosaccharides and, to a certain extent, lactose in the permeate stream [12, 111]. Bearing in mind that saccharides are neutral molecules, their separation is mainly based on the differences in molecular size and convective and diffusion phenomena [111]. Therefore, based on their molecular weights, nanofiltration proved to be the method of choice, although, membranes might not be selective enough since the separation of molecules of similar molecular weight is expected [11]. However, with genuine optimization of the most relevant parameters for nanofiltration - effective transmembrane pressure, temperature, flow, type of configuration, the molecular weight cut-off value (MWCO), the pore size distribution and the morphology of the membrane [108], the GOS purification can be successfully performed. The main constraints of this technique, in the beginning, were highly concentrated and viscous solutions and high osmotic pressure, which caused membrane fouling and, consequently, flux decay and concentration polarization effect [112]. Therefore, it was proposed that nanofiltration of GOS should be conducted using low solute concentrations (around 100 g/l). Since, GOS synthesis must be performed at concentrations higher than 300 g/l, the undesirable dilution of the GOS should be performed [12]. Goulas evaluated the purification of diluted commercial GOS using nanofiltration with a cross-flow filtration unit in an attempt to investigate the significance of the operating parameters on GOS purification [113], and concluded that the increase in effective transmembrane pressure provided a flux increase. However, this further resulted in high solute retention on the membrane surface and reduction of the pore size, thus increasing the oligosaccharides retention, disaccharides and monosaccharides retention, and membrane selectivity shift. In terms of concentrated GOS solutions fractionation, nanofiltration was performed at higher temperatures that allowed viscosity reduction, flux increase and membrane fouling reduction, as well as the microbial contamination repression [114]. However, this was obtained at the expense of membrane pore size increase, causing the significant portion of GOS permeation, while also reducing the performance of purification [114]. In contrast to this, the use of low temperatures caused the increase of GOS rejection coefficient, which then resulted in better GOS separation, although with significantly increased operation period, while retainig the reduction of the microbial contamination [106]. Nevertheless, the balance between purity and attained GOS yield had to be thoroughly examined and established. In order to overcome these drawbacks, certain multistage processes were proposed by several authors [115-116]. Over the last few years, the application of membrane cascades for the purification of complex mixtures has gained interest, because membranes of different properties can be combined, thus resulting in better purification performances. Also, solvent consumption can be lowered by recycling a solvent after its recovery through the use of extra filtration stages [116]. Several other alternative procedures to upgrade the purification performance of membrane separations have been proposed: pre-hydrolysis of raw GOS in terms of lowering the lactose content which allows easier separation of GOS from remaining monosaccharides, the usage of vibrating membrane modules, electrofiltration, and pulse feeding [11].

Efficient purification of obtained GOS by removal of monosaccharides and the remaining lactose without lowering GOS content is undoubtedly important from the application point of view. However, an effective downstream process might cause a sharp increase in process costs, and significantly influence the cost-effectiveness of the GOS production process. Therefore, thorough optimization and cost estimation of these techniques should be performed when determining their industrial potential.

COMMERCIAL PRODUCTION OF GOS

Nowadays, the production of GOS is well-established on an industrial scale and several GOS products are available on the global market. It is expected to reach USD 1.01 billion by 2020 [117]. It is interesting that a limited number of companies still hold a substantial amount of total GOS market. For example, the pioneer in the field of GOS production - Yakult Pharmaceutical industry Co. from Japan, still represents a leading GOS producer, while FrieslandCampina Domo is slowly becoming dominant in Europe. Some of the most important GOS producing companies are listed in Table 1. All these companies are focusing their research on development and enhancement of product portfolios, and keeping up with the emerging trends of the GOS market. GOS products that are currently present on the global market are most commonly found in powder or syrup form, and produced according to the similar production scheme (Figure 6).

Table 1. The most relevant commercial GOS products

Producer	Product	GOS (%)
FrieslandCampina Domo	Vivinal® GOS Syrup	59
	Vivinal® GOS Syrup Organic	59
	Vivinal® GOS Omni Syrup	63
	Vivinal® GOS Easy Drying Syrup	72
	Vivinal® GOS Powder	71
	Vivinal® GOS Powder WPC	30
	Vivinal® GOS Powder Maltodextrin	30
Yakult Pharmaceutical Industry Co., Ltd	Oligomate® 55N	≥55
	Oligomate® 55NP	≥55
Clasado BioSciences	Bimuno® (DAILY, IBAID, TRAVELAID)	65
Nissin Sugar Co., Ltd.	Cup Oligo H70	70
	Cup Oligo P	70
Dairy Crest Limited	NZMP SureStart™ GOS 57 Syrup	57
	NZMP SureStart™ Organic GOS 57 Syrup	57
	NZMP SureStart™ GOS 70 Syrup	70
	NZMP SureStart™ GOS 70 Powder	70
	Promovita® GOS 65 Syrup	65
	Promovita® GOS 70 Syrup	70
	Promovita® GOS 70 Powder	70
	Nutrabiotic® GOS 65 Syrup	65
	Nutrabiotic® GOS 70 Syrup	70
	Nutrabiotic® GOS 70 Powder	70
GTC Nutrition	Purimune™	≥90
Vitalus Nutrition Inc.	Vitagos™	66
Nestlé Nutrition	Nestlé GOS	50
New Francisco Biotechnology Corporation (NFBC)	King-Prebiotics® GOS-270-P	27
	King-Prebiotics® GOS-700-P	70
	King-Prebiotics® GOS-800-P	80
	King-Prebiotics® GOS-900-P	90
	King-Prebiotics® GOS-1000-P	100
	King-Prebiotics® GOS-570-S	57

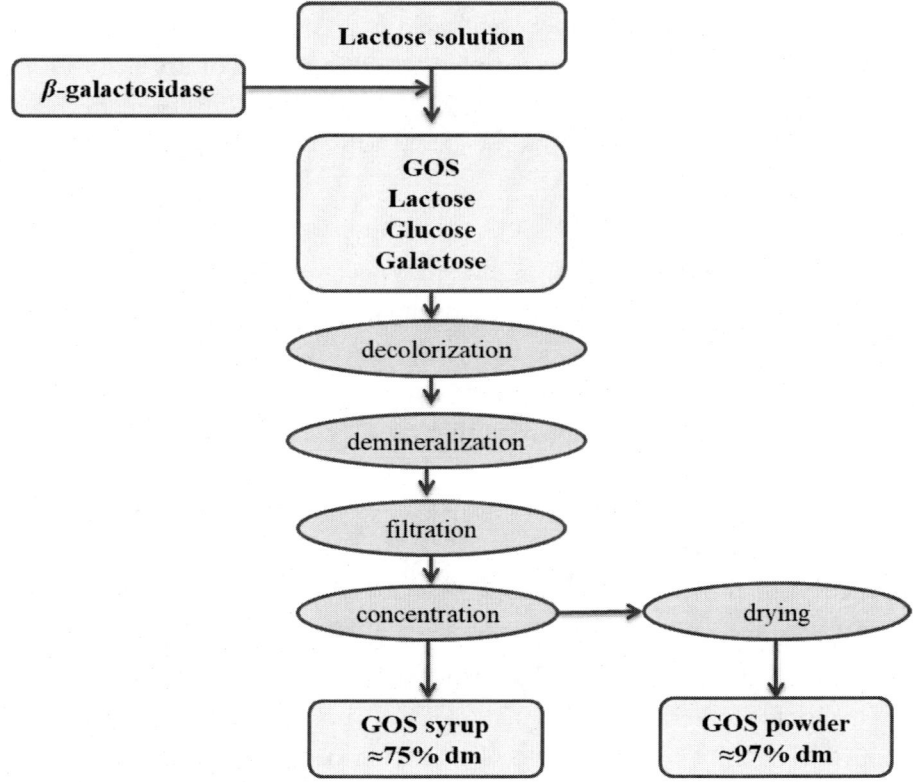

Figure 6. Typical GOS production process scheme (adapted from Lamsal, 2012) [6].

Firstly, the reaction substrate (lactose or lactose rich dairy by-product) is solubilized and pH and temperature are adjusted to fit the optimum reaction conditions for the employed enzyme. This kind of production usually involves the conversion of the concentrated solutions of lactose (20-40% w/w). After the addition of the chosen enzyme, the reaction takes place in batch or continuous reactors with stirring. When the optimum GOS concentration is reached, the reaction is interrupted (pH adjustment or thermal inactivation [29]) and the product is then subjected to various purification steps (Figure 6). Colours and saccharides other than GOS are mostly removed by adsorption using activated charcoal, while proteins are mainly separated by ultrafiltration and ionic adsorption (e.g., Celite and Perlite). In the following step demineralization is achieved by ion exchange chromatography using weak acid or basic resins [10]. The final product is then filtered and concentrated to 67-74% w/w of dry matter for the syrup production, or to a concentration of 50% w/w of dry matter for the production of powdered preparations, which are obtained through subsequent spray drying.

The obtained commercial GOS preparations (both syrup or powder form) are still complex mixtures of oligosaccharides of varying purity, since the majority of them still contain unreacted lactose and products of hydrolysis (glucose and glucose) [6]. The GOS concentration in commercial products varies between 20 and 90% of dry matter (Table 1). It is important, however, to emphasize that these products do not differ only in the GOS content, but also in the composition of GOS, namely the degree of polymerization and the type of β-linkages within them. As mentioned earlier, this is most likely a consequence of the nature of the applied enzyme. Namely, Aspergillus oryzae β-galactosidase catalyzed production of GOS yielded Oligomate® give products rich in β-(1→6) linked GOS, while β-(1→3) linkages were dominant within GOS of Bimuno products [118]. On the other hand, Vivinal® GOS, Cup Oligo and Purimune™ products generally contain galacto-oligosaccharides dominated by β-(1→4) glycoside bonds [1].

APPLICATION OF GOS

The aforementioned tremendous expected growth (CAGR of approximately 10% over the forecast period of 2018-2024) in GOS ingredient market is direct consequence of aroused public awareness about the beneficial effects of prebiotics on overall health. GOS are recognized as prebiotic compounds by the definition of Gibson and co-workers [119] claiming that a prebiotic is a selectively fermented ingredient that results in specific changes in the composition and/or activity of the gastrointestinal (GI) microbiota thus conferring benefit(s) upon host health.

Since they represent a complex group of carbohydrates consisting of glucose and galactose connected via β-glycosidic bonds, they are generally not subjected to degradation in the upper part of the GI tract, and thus arrive unchanged to the colon where they carry out their prebiotic function [120]. Their beneficial effect on the health of the host is observed in two basic mechanisms. Primarily, they allow selective proliferation of useful microbial bacteria (especially bifidobacteria and lactobacilli), which further protect the organism from the colonization of pathogenic species and improve the host's immune system, thereby preventing the occurrence of various inflammatory problems [121-122]. On the other hand, the fermentation of GOS leads to the short-chain fatty acids (SCFAs) synthesis in the colon [123]. Most of these acids are absorbed in the GI tract, allowing the host to use the energy from the food not digested in the upper parts of the digestive system. More importantly, their presence is associated with numerous positive effects on human health, such as: reduction of lumen pH, increasing mineral absorption, controlling the fat and carbohydrate metabolism, and cholesterol level in the serum, increase the frequency of defecation, and suppression of infections and inflammatory bowel diseases owing to their detoxification effect in the lumen [124-125].

In addition to these highly affirmative physiological properties, the widespread use of GOS is at the same time determined by their favourable physical/chemical properties [124]. Namely, GOS are completely soluble in water and other products, and more importantly, do not affect the viscosity of those products [7]. They are of neutral taste, slightly sweet (30-60% sweetness of sucrose), which is the reason why they do not change the taste of the product either [124]. They are exceptionally stable in acidic environments (at 37 °C for several months at pH 2), which is not the case with other prebiotics. Also, they are stable at extremely high temperatures (160 °C) for up to 10 minutes in neutral conditions, and for 10 minutes in acidic conditions (pH 2) at 100°C [7]. They are able to retain moisture, thus preventing the drying of the product. Additionally GOS improve the taste and texture of obtained products [124]. Since they are not degradable in the upper part of the digestive tract, they have a very low caloric value (1-2 kcal/g) and can be used in dietary products as a substitute for sugars. Also, being non-degradable for normal bacteria in the oral cavity made them the artificial sweetener of choice since they do not cause caries [6].

In recent years, GOS fields of applications have started widening. Although primarily used as ingredients in infant formulas due to the fact that they are able to mimic the beneficial effects of the oligosaccharides present in human breast milk, build and maintain the immune system, and protect against colonization of the pathogen during this sensitive period [3], thanks to their aforementioned significant prebiotic and good physical properties, GOS have recently been introduced to various products: foods and beverages (fermented dairy products, juices, pastries, jams, confectionary products), pet food and animal feed [126-128], and finally pharmaceutical and care products [5, 129]. The main advantages of these innovative products are superior nutritional and/or functional properties in comparison to the existing products, primarily due to their important prebiotic activities. Companies that want to be proactive and implement the results of the up-to-date health findings and recommendations and provide their customers the high quality value-added products are, therefore, very eager to include GOS in the wide range of their products.

Acknowledgments

The authors are grateful for the support of Ministry of Education, Science and Technological Development of Republic of Serbia through funding Projects III 46010 and TR 31035.

References

[1] Torres, D. P.; Gonçalves, M.; Teixeira, J. A.; Rodrigues, L. R. 2010. "Galacto-Oligosaccharides: Production, properties, applications, and significance as prebiotics." *Review of Comprehensive Reviews in Food Science and Food Safety* no. 9 (5):438-454. doi: 10.1111/j.1541-4337.2010.00119.x.

[2] Gosling, A.; Stevens, G. W.; Barber, A. R.; Kentish, S. E.; Gras, S. L. 2010. "Recent advances refining galactooligosaccharide production from lactose." *Review of Food Chemistry* no. 121 (2):307-318. doi: 10.1016/j.foodchem.2009.12.063.

[3] Macfarlane, G. T.; Steed, H.; Macfarlane, S. 2008. "Bacterial metabolism and health-related effects of galacto-oligosaccharides and other prebiotics." *Review of Journal of Applied Microbiology* no. 104 (2):305-344. doi:10.1111/j.1365-2672.2007.03520.x.

[4] Gibson, G. R.; Rastall, R. A. 2006. *Prebiotics: development & application.* Translated by. Edited by. ed. vols. Vol. 466. Wiley Online Library. Original edition. Reprint.

[5] Sangwan, V.; Tomar, S. K.; Singh, R. R. B.; Singh, A. K.; Ali, B. 2011. "Galactooligosaccharides: Novel Components of Designer Foods." *Review of Journal of Food Science* no. 76 (4):R103-R111. doi: 10.1111/j.1750-3841.2011.02131.x.

[6] Lamsal, B. P. 2012. "Production, health aspects and potential food uses of dairy prebiotic galactooligosaccharides." *Review of Journal of the Science of Food and Agriculture* no. 92 (10):2020-2028. doi: 10.1002/jsfa.5712.

[7] Van Leusen, E.; Torringa, E.; Groenink, P.; Kortleve, P.; Geene, R.; Schoterman, M.; Klarenbeek, B. 2014. "Industrial Applications of Galactooligosaccharides." In *Food Oligosaccharides: Production, Analysis and Bioactivity,* edited by, 470-491. Original edition.

[8] Sangwan, V.; Tomar, S. K.; Ali, B.; Singh, R. R. B.; Singh, A. K. 2015. "Galactooligosaccharides reduce infection caused by Listeria monocytogenes and modulate IgG and IgA levels in mice." *Review of International Dairy Journal* no. 41:58-63. doi: 10.1016/j.idairyj.2014.09.010.

[9] Wallenfels, K. 1951. "Enzymatische Synthese von Oligosacchariden aus Disacchariden." ["Enzymatic Synthesis of Oligosaccharides from Disaccharides"] *Review of Die Naturwissenschaften* no. 38 (13):306-307. doi: 10.1007/BF00636782.

[10] Panesar, P. S.; Kaur, R.; Singh, R. S.; Kennedy, J. F. 2018. "Biocatalytic strategies in the production of galacto-oligosaccharides

and its global status." *Review of International Journal of Biological Macromolecules* no. 111:667-679. doi: 10.1016/j.ijbiomac.2018. 01.062.

[11] Illanes, A.; Vera, C.; Wilson, L. 2016. "Enzymatic production of galacto-oligosaccharides." In *Lactose-Derived Prebiotics: A Process Perspective*, edited by, 111-189. Original edition.

[12] Vera, C.; Córdova, A.; Aburto, C.; Guerrero, C.; Suárez, S.; Illanes, A. 2016. "Synthesis and purification of galacto-oligosaccharides: state of the art." *Review of World Journal of Microbiology and Biotechnology* no. 32 (12):197. doi.

[13] Stahl, B.; Boehm, G. n. 2007. "Oligosaccharides from Milk." *Review of The Journal of Nutrition* no. 137 (3):847S-849S. doi: 10.1093/jn/137.3.847S.

[14] Huh, K. T.; Toba, T.; Adachi, S. 1990. "Oligosaccharide formation during the hydrolysis of lactose with hydrochloric acid and cation exchange resin." *Review of Food Chemistry* no. 38 (4):305-314. doi: 10.1016/0308-8146(90)90188-A.

[15] Huh, K. T.; Toba, T.; Adachi, S. 1991. "Oligosaccharide structures formed during acid hydrolysis of lactose." *Review of Food Chemistry* no. 39 (1):39-49. doi: 10.1016/0308-8146(91)90083-Z.

[16] Cruz, R.; Cruz, V. c. D. A.; Belote, J. G.; de Oliveira Khenayfes, M.; Dorta, C.; dos Santos Oliveira, L. z. H.; Ardiles, E.; Galli, A. 1999. "Production of transgalactosylated oligosaccharides (TOS) by galactosyltransferase activity from *Penicillium simplicissimum*." *Review of Bioresource technology* no. 70 (2):165-171. doi.

[17] Tzortzis, G.; Goulas, A. K.; Gee, J. M.; Gibson, G. R. 2005. "A novel galactooligosaccharide mixture increases the bifidobacterial population numbers in a continuous in vitro fermentation system and in the proximal colonic contents of pigs *in vivo*." *Review of The Journal of Nutrition* no. 135 (7):1726-1731. doi.

[18] Withers, S. G.; Lougheed, B. 1999. *Oligosaccharide synthesis using activated glycoside derivative, glycosyl transferase and catalytic amount of nucleotide phosphate*. In edited by: Google Patents. Original edition.

[19] Fischer, C.; Kleinschmidt, T. 2015. "Synthesis of galactooligosaccharides using sweet and acid whey as a substrate." *Review of International Dairy Journal* no. 48:15-22. doi: https://doi.org/10.1016/j.idairyj.2015.01.003.

[20] Nath, A.; Verasztó, B.; Basak, S.; Koris, A.; Kovács, Z.; Vatai, G. 2016. "Synthesis of Lactose-Derived Nutraceuticals from Dairy Waste Whey—a Review." *Review of Food and Bioprocess Technology* no. 9 (1):16-48. doi: 10.1007/s11947-015-1572-2.

[21] Nath, A.; Verasztó, B.; Basak, S.; Koris, A.; Kovács, Z.; Vatai, G. 2015. "Synthesis of Lactose-Derived Nutraceuticals from Dairy Waste Whey—a Review." *Review of Food and Bioprocess Technology* no. 9 (1):16-48. doi: 10.1007/s11947-015-1572-2.

[22] Pereira-Rodríguez, Á.; Fernández-Leiro, R.; González-Siso, M. I.; Cerdán, M. E.; Becerra, M.; Sanz-Aparicio, J. 2012. "Structural basis of specificity in tetrameric *Kluyveromyces lactis* β-galactosidase." *Review of Journal of Structural Biology* no. 177 (2):392-401. doi: http://dx.doi.org/10.1016/j.jsb.2011.11.031.

[23] Maksimainen, M. M.; Lampio, A.; Mertanen, M.; Turunen, O.; Rouvinen, J. 2013. "The crystal structure of acidic β-galactosidase from *Aspergillus oryzae.*" *Review of International Journal of Biological Macromolecules* no. 60:109-115. doi: http://dx.doi.org/10.1016/j.ijbiomac.2013.05.003.

[24] Juers, D. H.; Heightman, T. D.; Vasella, A.; McCarter, J. D.; Mackenzie, L.; Withers, S. G.; Matthews, B. W. 2001. "A structural view of the action of *Escherichia coli* (lac Z) β-galactosidase." *Review of Biochemistry* no. 40 (49):14781-14794. doi.

[25] Gloster, T. M.; Roberts, S.; Ducros, V. M. A.; Perugino, G.; Rossi, M.; Hoos, R.; Moracci, M.; Vasella, A.; Davies, G. J. 2004. "Structural studies of the β-glycosidase from *Sulfolobus solfataricus* in complex with covalently and noncovalently bound inhibitors." *Review of Biochemistry* no. 43 (20):6101-6109. doi: 10.1021/bi049666m.

[26] Otieno, D. O. 2010. "Synthesis of β-Galactooligosaccharides from Lactose Using Microbial β-Galactosidases." *Review of*

Comprehensive Reviews in Food Science and Food Safety no. 9 (5): 471-482. doi: 10.1111/j.1541-4337.2010.00121.x.

[27] Mahoney, R. R. 1998. "Galactosyl-oligosaccharide formation during lactose hydrolysis: a review." *Review of Food chemistry* no. 63 (2):147-154. doi.

[28] Carević, M.; Bezbradica, D.; Banjanac, K.; Milivojević, A.; Fanuel, M.; Rogniaux, H.; Ropartz, D.; Veličković, D. 2016. "Structural Elucidation of Enzymatically Synthesized Galacto-oligosaccharides Using Ion-Mobility Spectrometry-Tandem Mass Spectrometry." *Review of Journal of Agricultural and Food Chemistry* no. 64 (18):3609-3615. doi: 10.1021/acs.jafc.6b01293.

[29] Scott, F.; Vera, C.; Conejeros, R. 2016. "Technical and economic analysis of industrial production of lactose-derived prebiotics with focus on galacto-oligosaccharides." In *Lactose-Derived Prebiotics: A Process Perspective*, edited by, 261-284. Original edition.

[30] Vera, C.; Guerrero, C.; Conejeros, R.; Illanes, A. 2012. "Synthesis of galacto-oligosaccharides by β-galactosidase from Aspergillus oryzae using partially dissolved and supersaturated solution of lactose." *Review of Enzyme and Microbial Technology* no. 50 (3):188-194. doi.

[31] Huerta, L. M.; Vera, C.; Guerrero, C.; Wilson, L.; Illanes, A. 2011. "Synthesis of galacto-oligosaccharides at very high lactose concentrations with immobilized β-galactosidases from Aspergillus oryzae." *Review of Process Biochemistry* no. 46 (1):245-252. doi: 10.1016/j.procbio.2010.08.018.

[32] Park, H.-Y.; Kim, H.-J.; Lee, J.-K.; Kim, D.; Oh, D.-K. 2008. "Galactooligosaccharide production by a thermostable β-galactosidase from *Sulfolobus solfataricus*." *Review of World Journal of Microbiology and Biotechnology* no. 24 (8):1553-1558. doi.

[33] Park, A.-R.; Oh, D.-K. 2010. "Galacto-oligosaccharide production using microbial β-galactosidase: current state and perspectives." *Review of Applied Microbiology and Biotechnology* no. 85 (5):1279-1286. doi.

[34] Klein, M. P.; Fallavena, L. P.; Schöffer, J. d. N.; Ayub, M. A.; Rodrigues, R. C.; Ninow, J. L.; Hertz, P. F. 2013. "High stability of immobilized β-D-galactosidase for lactose hydrolysis and galactooligosaccharides synthesis." *Review of Carbohydrate polymers* no. 95 (1):465-470. doi.

[35] Albayrak, N.; Yang, S. T. 2002. "Production of galacto-oligosaccharides from lactose by Aspergillus oryzae β-galactosidase immobilized on cotton cloth." *Review of Biotechnology and Bioengineering* no. 77 (1):8-19. doi.

[36] Carević, M.; Ćorović, M.; Mihailović, M.; Banjanac, K.; Milisavljević, A.; Veličković, D.; Bezbradica, D. 2016. "Galacto-oligosaccharide synthesis using chemically modified β-galactosidase from *Aspergillus oryzae* immobilised onto macroporous amino resin." *Review of International Dairy Journal* no. 54:50-57. doi.

[37] Nath, A.; Mondal, S.; Chakraborty, S.; Bhattacharjee, C.; Chowdhury, R. 2014. "Production, purification, characterization, immobilization, and application of β-galactosidase: A review." *Review of Asia-Pacific Journal of Chemical Engineering* no. 9 (3):330-348. doi: 10.1002/apj.1801.

[38] Boon, M. A.; Janssen, A. E. M.; Van't Riet, K. 2000. "Effect of temperature and enzyme origin on the enzymatic synthesis of oligosaccharides." *Review of Enzyme and Microbial Technology* no. 26 (2-4):271-281. doi: 10.1016/S0141-0229(99)00167-2.

[39] Chen, X. Y.; Gänzle, M. G. 2017. "Lactose and lactose-derived oligosaccharides: More than prebiotics?" *Review of International Dairy Journal* no. 67:61-72. doi.

[40] Hansson, T.; Kaper, T.; Van Oost, J. D.; De Vos, W. M.; Adlercreutz, P. 2001. "Improved oligosaccharide synthesis by protein engineering of β-glucosidase CelB from hyperthermophilic *Pyrococcus furiosus*." *Review of Biotechnology and Bioengineering* no. 73 (3):203-210. doi: 10.1002/bit.1052.

[41] Jørgensen, F.; Hansen, O. C.; Stougaard, P. 2001. "High-efficiency synthesis of oligosaccharides with a truncated β-galactosidase from Bifidobacterium bifidum." *Review of Applied Microbiology and*

Biotechnology no. 57 (5-6):647-652. doi: 10.1007/s00253-001-0845-z.

[42] Gänzle, M. G. 2012. "Enzymatic synthesis of galactooligosaccharides and other lactose derivatives (hetero-oligosaccharides) from lactose." *Review of International Dairy Journal* no. 22 (2):116-122. doi: 10.1016/j.idairyj.2011.06.010.

[43] Vera, C.; Guerrero, C.; Illanes, A. 2011. "Determination of the transgalactosylation activity of *Aspergillus oryzae* β-galactosidase: effect of pH, temperature, and galactose and glucose concentrations." *Review of Carbohydrate Research* no. 346 (6):745-752. doi:

[44] Park, H.-Y.; Kim, H.-J.; Lee, J.-K.; Kim, D.; Oh, D.-K. 2007. "Galactooligosaccharide production by a thermostable β-galactosidase from *Sulfolobus solfataricus*." *Review of World Journal of Microbiology and Biotechnology* no. 24 (8):1553-1558. doi: 10.1007/s11274-007-9642-x.

[45] Benavente, R.; Pessela, B. C.; Curiel, J. A.; De Las Rivas, B.; Muñoz, R.; Guisán, J. M.; Mancheño, J. M.; Cardelle-Cobas, A.; Ruiz-Matute, A. I.; Corzo, N. 2015. "Improving properties of a novel β-galactosidase from *Lactobacillus plantarum* by covalent immobilization." *Review of Molecules* no. 20 (5):7874-7889. doi: 10.3390/molecules20057874.

[46] Black, B. A.; Lee, V. S. Y.; Zhao, Y. Y.; Hu, Y.; Curtis, J. M.; Gänzle, M. G. 2012. "Structural identification of novel oligosaccharides produced by *Lactobacillus bulgaricus* and *Lactobacillus plantarum*." *Review of Journal of Agricultural and Food Chemistry* no. 60 (19):4886-4894. doi: 10.1021/jf300917m.

[47] Gobinath, D.; Prapulla, S. 2015. "Transgalactosylating β-galactosidase from probiotic Lactobacillus plantarum MCC2156: production and permeabilization for use as whole cell biocatalyst." *Review of J Food Sci Technol* no. 52 (9):6003-6009. doi: 10.1007/s13197-014-1656-4.

[48] Hsu, C. A.; Lee, S. L.; Chou, C. C. 2007. "Enzymatic production of galactooligosaccharides by β-galactosidase from *Bifidobacterium*

longum BCRC 15708." *Review of Journal of Agricultural and Food Chemistry* no. 55 (6):2225-2230. doi: 10.1021/jf063126+.
[49] Iqbal, S.; Nguyen, T. H.; Nguyen, H. A.; Nguyen, T. T.; Maischberger, T.; Kittl, R.; Haltrich, D. 2011. "Characterization of a heterodimeric GH2 β-galactosidase from lactobacillus sakei Lb790 and formation of prebiotic galacto-oligosaccharides." *Review of Journal of Agricultural and Food Chemistry* no. 59 (8):3803-3811. doi: 10.1021/jf103832q.
[50] Liu, G. X.; Kong, J.; Lu, W. W.; Kong, W. T.; Tian, H.; Tian, X. Y.; Huo, G. C. 2011. "β-Galactosidase with transgalactosylation activity from *Lactobacillus fermentum* K4." *Review of Journal of Dairy Science* no. 94 (12):5811-5820. doi: http://dx.doi.org/10.3168/jds.2011-4479.
[51] Splechtna, B.; Nguyen, T. H.; Steinböck, M.; Kulbe, K. D.; Lorenz, W.; Haltrich, D. 2006. "Production of prebiotic galacto-oligosaccharides from lactose using β-galactosidases from *Lactobacillus reuteri.*" *Review of Journal of Agricultural and Food Chemistry* no. 54 (14):4999-5006. doi: 10.1021/jf053127m.
[52] Tzortzis, G.; Goulas, A. K.; Gibson, G. R. 2005. "Synthesis of prebiotic galactooligosaccharides using whole cells of a novel strain, *Bifidobacterium bifidum* NCIMB 41171." *Review of Applied microbiology and biotechnology* no. 68 (3):412-416. doi.
[53] Carević, M.; Vukašinović-Sekulić, M.; Ćorović, M.; Rogniaux, H.; Ropartz, D.; Veličković, D.; Bezbradica, D. 2018. "Evaluation of β-galactosidase from *Lactobacillus acidophilus* as biocatalyst for galacto-oligosaccharides synthesis: Product structural characterization and enzyme immobilization." *Review of Journal of bioscience and bioengineering* no. 126 (6):697-704. doi.
[54] Albayrak, N.; Yang, S. T. 2002. "Production of galacto-oligosaccharides from lactose by *Aspergillus oryzae* beta-galactosidase immobilized on cotton cloth." *Review of Bio-technology and bioengineering* no. 77 (1):8-19. doi.
[55] Gaur, R.; Pant, H.; Jain, R.; Khare, S. K. 2006. "Galacto-oligosaccharide synthesis by immobilized Aspergillus oryzae β-

galactosidase." *Review of Food Chemistry* no. 97 (3):426-430. doi: 10.1016/j.foodchem.2005.05.020.
[56] Giacomini, C.; Villarino, A.; Franco-Fraguas, L.; Batista-Viera, F. 1998. "Immobilization of β-galactosidase from *Kluyveromyces lactis* on silica and agarose: Comparison of different methods." *Review of Journal of Molecular Catalysis - B Enzymatic* no. 4 (5-6):313-327. doi: 10.1016/S1381-1177(98)00071-X.
[57] Grosová, Z.; Rosenberg, M.; Rebroš, M.; Šipocz, M.; Sedláčková, B. 2008. "Entrapment of β-galactosidase in polyvinylalcohol hydrogel." *Review of Biotechnology Letters* no. 30 (4):763-767. doi: 10.1007/s10529-007-9606-0.
[58] Guidini, C. Z.; Fischer, J.; Santana, L. N. S.; Cardoso, V. L.; Ribeiro, E. J. 2010. "Immobilization of *Aspergillus oryzae* β-galactosidase in ion exchange resins by combined ionic-binding method and cross-linking." *Review of Biochemical Engineering Journal* no. 52 (2-3):137-143. doi: 10.1016/j.bej.2010.07.013.
[59] Klein, M. P.; Nunes, M. R.; Rodrigues, R. C.; Benvenutti, E. V.; Costa, T. M. H.; Hertz, P. F.; Ninow, J. L. 2012. "Effect of the support size on the properties of β-galactosidase immobilized on chitosan: Advantages and disadvantages of macro and nanoparticles." *Review of Biomacromolecules* no. 13 (8):2456-2464. doi: 10.1021/bm3006984.
[60] Neri, D. F. M.; Balcão, V. M.; Costa, R. S.; Rocha, I. C. A. P.; Ferreira, E. M. F. C.; Torres, D. P. M.; Rodrigues, L. R. M.; Carvalho Jr, L. B.; Teixeira, J. A. 2009. "Galacto-oligosaccharides production during lactose hydrolysis by free *Aspergillus oryzae* β-galactosidase and immobilized on magnetic polysiloxane-polyvinyl alcohol." *Review of Food Chemistry* no. 115 (1):92-99. doi: 10.1016/j.foodchem.2008.11.068.
[61] Torres, R.; Mateo, C.; Fernández-Lorente, G.; Ortiz, C.; Fuentes, M.; Palomo, J. M.; Guisan, J. M.; Fernández-Lafuente, R. 2003. "A novel heterofunctional epoxy-amino sepabeads for a new enzyme immobilization protocol: Immobilization-stabilization of β-

galactosidase from *Aspergillus oryzae*." *Review of Biotechnology Progress* no. 19 (3):1056-1060. doi: 10.1021/bp025771g.

[62] Urrutia, P.; Mateo, C.; Guisan, J. M.; Wilson, L.; Illanes, A. 2013. "Immobilization of Bacillus circulans β-galactosidase and its application in the synthesis of galacto-oligosaccharides under repeated-batch operation." *Review of Biochemical Engineering Journal* no. 77:41-48. doi: 10.1016/j.bej.2013.04.015.

[63] Rodriguez-Colinas, B.; de Abreu, M. A.; Fernandez-Arrojo, L.; de Beer, R.; Poveda, A.; Jimenez-Barbero, J.; Haltrich, D.; Ballesteros Olmo, A. O.; Fernandez-Lobato, M.; Plou, F. J. 2011. "Production of Galacto-oligosaccharides by the β-Galactosidase from *Kluyveromyces lactis*: Comparative Analysis of Permeabilized Cells versus Soluble Enzyme." *Review of Journal of Agricultural and Food Chemistry* no. 59 (19):10477-10484. doi: 10.1021/jf2022012.

[64] Yu, L.; O'Sullivan, D. J. 2014. "Production of galactooligosaccharides using a hyperthermophilic β-galactosidase in permeabilized whole cells of *Lactococcus lactis*." *Review of Journal of Dairy Science* no. 97 (2):694-703. doi: https://doi.org/10.3168/jds.2013-7492.

[65] Gosling, A.; Stevens, G. W.; Barber, A. R.; Kentish, S. E.; Gras, S. L. 2011. "Effect of the substrate concentration and water activity on the yield and rate of the transfer reaction of β-galactosidase from bacillus circulans." *Review of Journal of Agricultural and Food Chemistry* no. 59 (7):3366-3372. doi: 10.1021/jf104397w.

[66] Warmerdam, A.; Benjamins, E.; de Leeuw, T. F.; Broekhuis, T. A.; Boom, R. M.; Janssen, A. E. M. 2014. "Galacto-oligosaccharide production with immobilized β-galactosidase in a packed-bed reactor vs. free β-galactosidase in a batch reactor." *Review of Food and Bioproducts Processing* no. 92 (4):383-392. doi: http://dx.doi.org/10.1016/j.fbp.2013.08.014.

[67] Cardelle-Cobas, A.; Villamiel, M.; Olano, A.; Corzo, N. 2008. "Study of galacto-oligosaccharide formation from lactose using

Pectinex Ultra SP-L." *Review of Journal of the Science of Food and Agriculture* no. 88 (6):954-961. doi: 10.1002/jsfa.3173.

[68] Peterson, M. E.; Daniel, R. M.; Danson, M. J.; Eisenthal, R. 2007. "The dependence of enzyme activity on temperature: determination and validation of parameters." *Review of Biochemical Journal* no. 402 (2):331-337. doi.

[69] Iwasaki, K. I.; Nakajima, M.; Nakao, S. I. 1996. "Galacto-oligosaccharide production from lactose by an enzymic batch reaction using β-galactosidase." *Review of Process Biochemistry* no. 31 (1):69-76. doi.

[70] Placier, G.; Watzlawick, H.; Rabiller, C.; Mattes, R. 2009. "Evolved β-galactosidases from Geobacillus stearothermophilus with improved transgalactosylation yield for galacto-oligosaccharide production." *Review of Applied and environmental microbiology* no. 75 (19):6312-6321. doi.

[71] Oliveira, C.; Guimarães, P. M.; Domingues, L. 2011. "Recombinant microbial systems for improved β-galactosidase production and biotechnological applications." *Review of Biotechnology advances* no. 29 (6):600-609. doi.

[72] Shin, H. J.; Yang, J. W. 1994. "Galacto-oligosaccharide production by β-galactosidase in hydrophobic organic media." *Review of Biotechnology Letters* no. 16 (11):1157-1162. doi: 10.1007/BF01020843.

[73] Wang, K.; Lu, Y.; Liang, W. Q.; Wang, S. D.; Jiang, Y.; Huang, R.; Liu, Y. H. 2012. "Enzymatic synthesis of galacto-oligosaccharides in an organic-aqueous biphasic system by a novel β-galactosidase from a metagenomic library." *Review of Journal of Agricultural and Food Chemistry* no. 60 (15):3940-3946. doi: 10.1021/jf300890d.

[74] Klewicki, R. 2007. "Effect of selected parameters of lactose hydrolysis in the presence of β-galactosidase from various sources on the synthesis of galactosyl-polyol derivatives." *Review of Engineering in Life Sciences* no. 7 (3):268-274. doi: 10.1002/elsc.200620185.

[75] Kaftzik, N.; Neumann, S.; Kula, M. R.; Kragl, U. 2003. Enzymatic condensation reactions in ionic liquids. In *ACS Symposium Series*, edited by. Original edition.

[76] Gobinath, D.; Prapulla, S. G. 2014. "Permeabilized probiotic Lactobacillus plantarum as a source of β-galactosidase for the synthesis of prebiotic galactooligosaccharides." *Review of Biotechnology Letters* no. 36 (1):153-157. doi: 10.1007/s10529-013-1345-9.

[77] Manera, A. P.; De Almeida Costa, F. A.; Rodrigues, M. I.; Kalil, S. J.; Maugeri Filho, F. 2010. "Galacto-oligosaccharides production using permeabilized cells of *Kluyveromyces marxianus*." *Review of International Journal of Food Engineering* no. 6 (6). doi: 10.2202/1556-3758.1917.

[78] Foda, M. I.; Lopez-Leiva, M. 2000. "Continuous production of oligosaccharides from whey using a membrane reactor." *Review of Process Biochemistry* no. 35 (6):581-587. doi: 10.1016/S0032-9592(99)00108-9.

[79] Pocedičová, K.; Čurda, L.; Mišún, D.; Dryáková, A.; Diblíková, L. 2010. "Preparation of galacto-oligosaccharides using membrane reactor." *Review of Journal of Food Engineering* no. 99 (4):479-484. doi: 10.1016/j.jfoodeng.2010.02.001.

[80] Banjanac, K.; Carević, M.; Ćorović, M.; Milivojević, A.; Prlainović, N.; Marinkovic, A.; Bezbradica, D. 2016. *Novel β-galactosidase nanobiocatalyst systems for application in the synthesis of bioactive galactosides.* Translated by. Edited by. ed. vols. Vol. 6. Original edition. Reprint.

[81] Gänzle, M. G.; Haase, G.; Jelen, P. 2008. "Lactose: Crystallization, hydrolysis and value-added derivatives." *Review of International Dairy Journal* no. 18 (7):685-694. doi: 10.1016/j.idairyj.2008.03.003.

[82] Sen, D.; Gosling, A.; Stevens, G. W.; Bhattacharya, P. K.; Barber, A. R.; Kentish, S. E.; Bhattacharjee, C.; Gras, S. L. 2011. "Galactosyl oligosaccharide purification by ethanol precipitation." *Review of*

Food Chemistry no. 128 (3):773-777. doi: 10.1016/j. food chem.2011.03.076.

[83] Montañés, F.; Corzo, N.; Olano, A.; Reglero, G.; Ibáñez, E.; Fornari, T. 2008. "Selective fractionation of carbohydrate complex mixtures by supercritical extraction with CO_2 and different co-solvents." *Review of Journal of Supercritical Fluids* no. 45 (2):189-194. doi: 10.1016/j.supflu.2007.08.012.

[84] Montañés, F.; Fornari, T.; Martín-Álvarez, P. J.; Montilla, A.; Corzo, N.; Olano, A.; Ibáñez, E. 2007. "Selective fractionation of disaccharide mixtures by supercritical CO_2 with ethanol as co-solvent." *Review of Journal of Supercritical Fluids* no. 41 (1):61-67. doi: 10.1016/j.supflu.2006.08.010.

[85] Nobre, C.; Teixeira, J. A.; Rodrigues, L. R. 2012. "Fructo-oligosaccharides purification from a fermentative broth using an activated charcoal column." *Review of New Biotechnology* no. 29 (3):395-401. doi: 10.1016/j.nbt.2011.11.006.

[86] Hernández, O.; Ruiz-Matute, A. I.; Olano, A.; Moreno, F. J.; Sanz, M. L. 2009. "Comparison of fractionation techniques to obtain prebiotic galactooligosaccharides." *Review of International Dairy Journal* no. 19 (9):531-536. doi: 10.1016/j.idairyj.2009.03.002.

[87] Moreno, F. J.; Luz Sanz, M. 2014. Food Oligosaccharides: Production, Analysis and Bioactivity. Translated by. Edited by. ed. vols. Vol. 9781118426494, *Food Oligosaccharides: Production, Analysis and Bioactivity*. Original edition. Reprint.

[88] Koizumi, K. 2002. Chapter 3 HPLC of carbohydrates on graphitized carbon columns. In *Journal of Chromatography Library,* edited by. Original edition.

[89] Searle, L. E. J.; Cooley, W. A.; Jones, G.; Nunez, A.; Crudgington, B.; Weyer, U.; Dugdale, A. H.; Tzortzis, G.; Collins, J. W.; Woodward, M. J.; La Ragione, R. M. 2010. "Purified galactooligosaccharide, derived from a mixture produced by the enzymic activity of *Bifidobacterium bifidum*, reduces *Salmonella enterica* serovar Typhimurium adhesion and invasion in vitro and in

vivo." *Review of Journal of Medical Microbiology* no. 59 (12):1428-1439. doi: 10.1099/jmm.0.022780-0.

[90] Shoaf, K.; Mulvey, G. L.; Armstrong, G. D.; Hutkins, R. W. 2006. "Prebiotic galactooligosaccharides reduce adherence of enteropathogenic Escherichia coli to tissue culture cells." *Review of Infection and Immunity* no. 74 (12):6920-6928. doi: 10.1128/IAI.01030-06.

[91] Nobre, C.; Teixeira, J. A.; Rodrigues, L. R. 2015. "New trends and technological challenges in the industrial production and purification of fructo-oligosaccharides." *Review of Critical reviews in food science and nutrition* no. 55 (10):1444-1455. doi.

[92] Moravčík, J.; Gramblička, M.; Wiśniewski, Ł.; Vaňková, K.; Polakovič, M. 2012. "Influence of the ionic form of a cation-exchange adsorbent on chromatographic separation of galactooligosaccharides." *Review of Chemical Papers* no. 66 (6):583-588. doi.

[93] Wiśniewski, Ł.; Pereira, C. S.; Polakovič, M.; Rodrigues, A. E. 2014. "Chromatographic separation of prebiotic oligosaccharides. Case study: separation of galacto-oligosaccharides on a cation exchanger." *Review of Adsorption* no. 20 (2-3):483-492. doi.

[94] Brereton, K. R.; Green, D. B. 2012. "Isolation of saccharides in dairy and soy products by solid-phase extraction coupled with analysis by ligand-exchange chromatography." *Review of Talanta* no. 100:384-390. doi: https://doi.org/10.1016/j.talanta.2012.08.003.

[95] Sanz-Valero, J. I. *Production of galacto-oligosaccharides from lactose by immobilized b-galactosidase and posterior chromatographic separation*. The Ohio State University, 2009.

[96] Rajendran, A.; Paredes, G.; Mazzotti, M. 2009. "Simulated moving bed chromatography for the separation of enantiomers." *Review of Journal of Chromatography A* no. 1216 (4):709-738. doi: https://doi.org/10.1016/j.chroma.2008.10.075.

[97] Imamoglu, S. 2002. "Simulated Moving Bed Chromatography (SMB) for Application in Bioseparation." In *Modern Advances in*

Chromatography, edited by Freitag, R., 211-231. Berlin, Heidelberg: Springer Berlin Heidelberg. Original edition.
[98] Kovács, Z.; Benjamins, E.; Grau, K.; Rehman, A. U.; Ebrahimi, M.; Czermak, P. 2013. "Recent developments in manufacturing oligosaccharides with prebiotic functions." In *Biotechnology of food and feed additives,* edited by, 257-295. Springer. Original edition.
[99] Wiśniewski, Ł.; Antošová, M.; Polakovič, M. 2013. "Simulated moving bed chromatography separation of galactooligosaccharides." *Review of Acta Chimica Slovaca* no. 6 (2):206-210. doi.
[100] Geisser, A.; Hendrich, T.; Boehm, G.; Stahl, B. 2005. "Separation of lactose from human milk oligosaccharides with simulated moving bed chromatography." *Review of Journal of Chromatography A* no. 1092 (1):17-23. doi.
[101] Li, Z.; Xiao, M.; Lu, L.; Li, Y. 2008. "Production of non-monosaccharide and high-purity galactooligosaccharides by immobilized enzyme catalysis and fermentation with immobilized yeast cells." *Review of Process Biochemistry* no. 43 (8):896-899. doi.
[102] Goulas, A.; Tzortzis, G.; Gibson, G. R. 2007. "Development of a process for the production and purification of α-and β-galactooligosaccharides from Bifidobacterium bifidum NCIMB 41171." *Review of International Dairy Journal* no. 17 (6):648-656. doi.
[103] Santibáñez, L.; Guerrero, C.; Illanes, A. 2017. "Raw galactooligosaccharide purification by consecutive lactose hydrolysis and selective bioconversion." *Review of International Dairy Journal* no. 75:91-100. doi: https://doi.org/10.1016/j.idairyj.2017.07.008.
[104] Guerrero, C.; Vera, C.; Illanes, A. 2018. "Selective bioconversion with yeast for the purification of raw lactulose and transgalactosylated oligosaccharides." *Review of International Dairy Journal* no. 81:131-137. doi: https://doi.org/10.1016/j.idairyj. 2018.02.003.
[105] Aburto, C.; Guerrero, C.; Vera, C.; Wilson, L.; Illanes, A. 2016. "Simultaneous synthesis and purification (SSP) of galacto-

oligosaccharides in batch operation." *Review of LWT-Food Science and Technology* no. 72:81-89. doi.

[106] Conidi, C.; Drioli, E.; Cassano, A. 2018. "Membrane-based agro-food production processes for polyphenol separation, purification and concentration." *Review of Current Opinion in Food Science* no. 23:149-164. doi: 10.1016/j.cofs.2017.10.009.

[107] Castro-Muñoz, R.; Fíla, V. 2018. "Membrane-based technologies as an emerging tool for separating high-added-value compounds from natural products." *Review of Trends in Food Science and Technology* no. 82:8-20. doi: 10.1016/j.tifs.2018.09.017.

[108] Córdova, A.; Astudillo, C.; Illanes, A. 2019. "Membrane Technology for the Purification of Enzymatically Produced Oligosaccharides." In *Separation of Functional Molecules in Food by Membrane Technology,* edited by, 113-153. Elsevier. Original edition.

[109] Li, J.; Chase, H. A. 2010. "Applications of membrane techniques for purification of natural products." *Review of Biotechnology letters* no. 32 (5):601-608. doi.

[110] Pinelo, M.; Jonsson, G.; Meyer, A. S. 2009. "Membrane technology for purification of enzymatically produced oligosaccharides: molecular and operational features affecting performance." *Review of Separation and Purification Technology* no. 70 (1):1-11. doi.

[111] Goulas, A. K.; Grandison, A. S.; Rastall, R. A. 2003. "Fractionation of oligosaccharides by nanofiltration." *Review of Journal of the Science of Food and Agriculture* no. 83 (7):675-680. doi.

[112] Córdova, A.; Astudillo, C.; Giorno, L.; Guerrero, C.; Conidi, C.; Illanes, A.; Cassano, A. 2016. "Nanofiltration potential for the purification of highly concentrated enzymatically produced oligosaccharides." *Review of Food and Bioproducts Processing* no. 98:50-61. doi.

[113] Goulas, A. K.; Kapasakalidis, P. G.; Sinclair, H. R.; Rastall, R. A.; Grandison, A. S. 2002. "Purification of oligosaccharides by nanofiltration." *Review of Journal of Membrane Science* no. 209 (1):321-335. doi.

[114] Pruksasri, S.; Nguyen, T.-H.; Haltrich, D.; Novalin, S. 2015. "Fractionation of a galacto-oligosaccharides solution at low and high temperature using nanofiltration." *Review of Separation and purification technology* no. 151:124-130. doi.

[115] Patil, N. V.; Schotel, T.; Rodríguez Gómez, C. V.; Aguirre Montesdeoca, V.; Sewalt, J. J.; Janssen, A. E.; Boom, R. M. 2016. "Continuous purification of galacto-oligosaccharide mixtures by using cascaded membrane filtration." *Review of Journal of Chemical Technology & Biotechnology* no. 91 (5):1478-1484. doi.

[116] Montesdeoca, V. A.; Van der Padt, A.; Boom, R.; Janssen, A. E. 2016. "Modelling of membrane cascades for the purification of oligosaccharides." *Review of Journal of Membrane Science* no. 520:712-722. doi.

[117] Kumar, S. 2015. *Galacto Oligosaccharides (GOS) Market Share, Trends to* 2020. Translated by. Edited by. ed. vols. Vol. Original edition. Reprint.

[118] Charalampopoulos, D.; Rastall, R. A. 2009. *Prebiotics and probiotics science and technology*. Translated by. Edited by. ed. vols. Vol. 1. Springer. Original edition. Reprint.

[119] Gibson, G. R.; Scott, K. P.; Rastall, R. A.; Tuohy, K. M.; Hotchkiss, A.; Dubert-Ferrandon, A.; Gareau, M.; Murphy, E. F.; Saulnier, D.; Loh, G. 2010. "Dietary prebiotics: current status and new definition." *Review of Food Sci Technol Bull Funct Foods* no. 7 (1):1-19. doi.

[120] Bruno-Barcena, J. M.; Azcarate-Peril, M. A. 2015. "Galacto-oligosaccharides and colorectal cancer: Feeding our intestinal probiome." *Review of Journal of Functional Foods* no. 12:92-108. doi: 10.1016/j.jff.2014.10.029.

[121] Grimaldi, R.; Swann, J. R.; Vulevic, J.; Gibson, G. R.; Costabile, A. 2016. "Fermentation properties and potential prebiotic activity of Bimuno® galacto-oligosaccharide (65% galacto-oligosaccharide content) on in vitro gut microbiota parameters." *Review of British Journal of Nutrition* no. 116 (3):480-486. doi: 10.1017/S0007 114516002269.

[122] So, D.; Whelan, K.; Rossi, M.; Morrison, M.; Holtmann, G.; Kelly, J. T.; Shanahan, E. R.; Staudacher, H. M.; Campbell, K. L. 2018. "Dietary fiber intervention on gut microbiota composition in healthy adults: A systematic review and meta-analysis." *Review of American Journal of Clinical Nutrition* no. 107 (6):965-983. doi: 10.1093/ajcn/nqy041.

[123] Carlson, J. L.; Erickson, J. M.; Lloyd, B. B.; Slavin, J. L. 2018. "Health effects and sources of prebiotic dietary fiber." *Review of Current developments in nutrition* no. 2 (3):nzy005. doi.

[124] Singla, V.; Chakkaravarthi, S. 2017. "Applications of prebiotics in food industry: A review." *Review of Food Science and Technology International* no. 23 (8):649-667. doi: 10.1177/1082013217721769.

[125] S, P. I.; P, P. 2014. "Prebiotics: Application in Bakery and Pasta Products." *Review of Critical Reviews in Food Science and Nutrition* no. 54 (4):511-522. doi: 10.1080/10408398.2011.590244.

[126] Biggs, P.; Parsons, C. M.; Fahey, G. C. 2007. "The effects of several oligosaccharides on growth performance, nutrient digestibilities, and cecal microbial populations in young chicks." *Review of Poultry Science* no. 86 (11):2327-2336. doi: 10.3382/ps.2007-00427.

[127] Jung, S. J.; Houde, R.; Baurhoo, B.; Zhao, X.; Lee, B. H. 2008. "Effects of galacto-oligosaccharides and a *Bifidobacteria lactis*-based probiotic strain on the growth performance and fecal microflora of broiler chickens." *Review of Poultry Science* no. 87 (9):1694-1699. doi: 10.3382/ps.2007-00489.

[128] Varasteh, S.; Braber, S.; Akbari, P.; Garssen, J.; Fink-Gremmels, J. 2015. "Differences in susceptibility to heat stress along the chicken intestine and the protective effects of galacto- oligosaccharides." *Review of PLoS ONE* no. 10 (9). doi: 10.1371/journal.pone.0138975.

[129] Krutmann, J. 2009. "Pre- and probiotics for human skin." *Review of Journal of Dermatological Science* no. 54 (1):1-5. doi: http://dx.doi.org/10.1016/j.jdermsci.2009.01.002.

BIOGRAPHICAL SKETCH

Milica Simović

Affiliation: Department of Biochemical Engineering and Biotechnology, Faculty of Technology and Metallurgy, University of Belgrade, Serbia

Education:
- Department of Biochemical Engineering and Biotechnology, Faculty of Technology and Metallurgy, University of Belgrade, Serbia (2005-2009), Dipl.-Ing. in Biochemical Engineering and Biotechnology
- Department of Biochemical Engineering and Biotechnology, Faculty of Technology and Metallurgy, University of Belgrade, Serbia (2009-2016), Ph.D. Engineering Technology/Biotechnology

Research and Professional Experience:
- Research trainee, Faculty of Technology and Metallurgy, University of Belgrade, Serbia (2011-2013)
- Research assistant, Faculty of Technology and Metallurgy, University of Belgrade, Serbia (2013-2017)
- Research associate, Faculty of Technology and Metallurgy, University of Belgrade, Serbia (2017 – today)
- Teaching assistant, Faculty of Technology and Metallurgy, University of Belgrade, Serbia (2012-today)

Scientific work of Dr. Milica Simović is mainly focused on the production and immobilization of different microbial enzymes (primarily galactosidases), and their application in the production of bioactive food components and cosmetic formulations in various reactor configurations. She defended her Doctoral Dissertation entitled "Production and immobilization of microbial β-galactosidases for implementation in transgalactosylation reactions" on September 5th, 2016 and earned her PhD degree in the field of Engineering Technology/Biotechnology.

During the work within the current national project she has been involved in various areas of biotechnology: optimization of enzymes (cellulase, protease, alpha galactosidase) microbial production, immobilization of cells and enzymes, bio-catalyzed synthesis of bioactive esters, and most recently valorization of different agro-industrial wastes.

During 2009 she was on research stay at the Institute of Industrial Lactology (INLAIN, UNL-CONICET), Faculty of Chemical Engineering, National University of the Littoral, Santa Fe, Argentina during three months.

Professional Appointments:
- Development of novel encapsulation and enzymatic technologies for production of biocatalysts and bioactive food components with purpose of improvement of their market competitiveness, quality and safety, Project No. 46010 within Program of Integrated Interdisciplinary Research of Ministry of Science and Technological Development of Republic of Serbia, 2011-today.
- Development of the procedure for fermentative production of phytopathogenic bacteria for application in biofungicides, Cooperation between Biogenesis Ltd. and Innovative Centre, faculty of Technology and Metallurgy, 2016.
- High protein soybean-based probiotic feed with increased digestibility, Innovation Fund CGS Program, 2017-2018.

Honors:
- "Panta S. Tutundžić" award for outstanding achievements during the studies (2006, 2007, 2008, and 2009)
- Recognition of the Serbian Chemical Society for exceptional success during the studies (2009).

Publications from the Last 3 Years:
1) Carević, M.: *Production and immobilization of microbial β-galactosidases for implementation in transgalactosylation reactions,*

Faculty of Technology and Metallurgy, Belgrade, 5th of September 2016.
2) Carević, M., Vukašinović-Sekulić, M., Ćorović, M., Rogniaux, H., Ropartz, D., Veličković D. and Bezbradica, D. (2018). Evaluation of β-galactosidase from *Lactobacillus acidophilus* as biocatalyst for galacto-oligosaccharides synthesis: Product structural characterization and enzyme immobilization. *Journal of Bioscience and Bioengineering*, 126 (VI): 697-704.
3) Mihajlovski, K., Radovanović, T., Carević, M. and Dimitrijević-Branković, S. (2018). Valorization of damaged rice grains: Optimization of bioethanol production by waste brewer's yeast using an amylolytic potential from the *Paenibacillus chitinolyticus* CKS1. *Fuel*, 224 (I): 591-599.
4) Ćorović, M., Milivojević, A., Carević, M., Banjanac, K., Jakovetić-Tanasković, S. And Bezbradica, D. (2017). Batch and semicontinuous production of L-ascorbyl oleate catalyzed by CALB immobilized onto Purolite® MN102. *Chemical Engineering Research & Design*, 26: 161-171.
5) Milivojević, A., Ćorović, M., Carević, M., Banjanac, K., Vujisić, Lj., Veličković, D., and Bezbradica, D. (2017). Highly efficient enzymatic acetylation of flavonoids: Development of solvent-free process and kinetic evaluation. *Biochemical Engineering Journal*, 128: 106-115.
6) Ćorović, M., Mihailović, M., Banjanac, K., Carević, M., Milivojević, A., Milosavić, N. And Bezbradica, D. (2017). Immobilization of Candida antarctica lipase B onto Purolite® MN102 and its application in solvent-free and organic media esterification. *Bioprocess and Biosystems Engineering*, 40 (I): 23-34.
7) Bezbradica, D., Ćorović, M., Jakovetić Tanasković, S., Luković, N., Carević, M., Milivojević, A. and Knezević-Jugović, Z. (2017). Enzymatic Syntheses of Esters-Green Chemistry for Valuable Food, Fuel and Fine Chemicals. *Current Organic Chemistry*, 21 (II):104-138.
8) Carević, M., Bezbradica, D., Banjanac, K., Milivojević, A., Fanuel, M., Rogniaux, H., Ropartz, D. and Veličković, D. (2016). Structural Elucidation of Enzymatically Synthesized Galacto-oligosaccharides

Using Ion-Mobility Spectrometry-Tandem Mass Spectrometry. *Journal of Agricultural and Food Chemistry* 64(18): 3609-3615.
9) Carević, M., Ćorović, M., Mihailović, M., Banjanac, K., Milisavljević, A., Veličković, D. and Bezbradica, D. (2016). Galacto-oligosaccharide synthesis using chemically modified β-galactosidase from *Aspergillus oryzae* immobilised onto macroporous amino resin. *International Dairy Journal*, 54: 50-57.
10) Banjanac, K., Mihailović, M., Prlainović, N., Ćorović, M., Carević, M., Marinković, A. And Bezbradica, D. (2016) Epoxy-silanization - tool for improvement of silica nanoparticles as support for lipase immobilization with respect to esterification activity. *Journal of Chemical Technology and Biotechnology*, 91(X): 2654-2663.
11) Banjanac, K., Carević, M., Ćorovć, M., Milivojević, A., Prlainović, N., Marinković, A. And Bezbradica, D. (2016). Novel β-galactosidase nanobiocatalyst systems for application in the synthesis of bioactive galactosides. *RSC Advances*, 6 (XCIX): 97216 – 97225.
12) Mihajlovski, K., Davidović, S., Carević, M., Radovanović, N., Šiler-Marinković, S., Rajilić-Stojanović, M. and Dimitrijević-Branković, S. (2016). Carboxymethyl cellulase production from a *Paenibacillus* sp, *Chemical industry*, 70 (III): 329-338.
13) Mihailović, M., Trbojević-Ivić, J., Banjanac, K., Milosavić, N., Veličković, D., Carević, M. and Bezbradica, D. Immobilization of maltase from Saccharomyces cerevisiae on thiosulfonate supports. *Journal of the Serbian Chemical Society*, 81 (II): 1371-1382.
14) Mihajlovski, K., Davidović, S., Veljović, Đ., Carević, M., Lazić, V., and Dimitrijević-Branković S. (2016). Effective valorisation of barley bran for simultaneous cellulase and β- amylase production by *Paenibacillus chitinolyticus* CKS1: Statistical optimization and enzymes application. *Journal of the Serbian Chemical Society*, 82 (XI): 1223-1236.
15) Carević, M., Banjanac, K., Ćorović, M., Jakovetić, S., Milivojević, A., Vukašinović-Sekulić M. and Bezbradica, D. (2016). Selection of lactic acid bacteria strain for simultaneous production of α-and β-galactosidases. *Zaštita materijala*, 57 (II): 265-273.

16) Miljković, M., Davidović, S., Carević, M., Veljović, Đ., Mladenović, D., Rajilić-Stojanović, M. and Dimitrijević-Branković, S. (2016). Sugar Beet Pulp as *Leuconostoc mesenteroides* T3 Support for Enhanced Dextransucrase Production on Molasses. *Applied Biochemistry and Biotechnology*, 180 (V):1016-1027.
17) Banjanac, K., Mihailović, M., Prlainović, N., Stojanović, M., Carević, M., Marinković, A. and Bezbradica, D. (2016) Cyanuric chloride functionalized silica nanoparticles for covalent immobilization of lipase. *Journal of Chemical Technology and Biotechnology*, 91 (X): 439-448.

In: Beta-Galactosidase
Editor: Eloy Kras

ISBN: 978-1-53615-605-8
© 2019 Nova Science Publishers, Inc.

Chapter 4

SCREENING AND IDENTIFICATION OF β-GALACTOSIDASE PRODUCING MICROORGANISMS FROM YAK YOGHOURT IN CHINA'S GANNAN PASTURE

Weibing Zhang[*], *Yingying Cao, Kaiyong Wen, Lei Cao, Jiang Ma and Qiaoqiao Luo*
College of Food Science and Engineering,
Gansu Agricultural University, Lanzhou, China

ABSTRACT

Strains with β-Galactosidase activity were isolated from 25 Yak Yoghourt samples collected from Gannan pasturing area of Gansu Province. An efficient β-Galactosidase producing strain SYA2 screened out from 21 strains was identified as *Enterobacter* sp. by the means of morphological feature observation, physiological and biochemical

---
[*] Corresponding Author's E-mail: 45330301@qq.com.

characteristics measurement, and 16S rDNA sequence analysis. The β-Galactosidase produced by *Enterobacter sp.* SYA2 was purified by ammonium sulfate precipitation. The optimum temperature, thermostability, optimum pH, and pH stability of the β-Galactosidase, as well as the effects of metal ions on enzymatic activity, were investigated. The optimum temperature of the β-Galactosidase was 40°C. The β-Galactosidase was stable at 35°C and 40°C, but the loss of activity was obvious at 45°C and 50°C. The optimum pH of the β-Galactosidase was 6.5. The β-Galactosidase was stable at pH 6.0~9.0. The β-Galactosidase activity could be activated by Mg^{2+}, Mn^{2+} and Na^+, while inhibited differently by Zn^{2+}, Cu^{2+} and Fe^{2+}. These results also indicate that SYA2 has a more notable effect on lactose hydrolysis in milk.

Keywords: screening, β-galactosidase, *Enterobacter* sp. SYA2

1. INTRODUCTION

β-Galactosidases (EC 3.2.1.23), which are also referred to as lactase, can hydrolyze galactosyl β(1→4) glucose of lactose into glucose and galactose and have transgalactosylation activity (Panesar et al., 2006). β-Galactosidases are widely used in dairy industries, pharmaceutical, analysis, and environmental protection, and occur in animal tissues, higher order plants, and microorganisms (Panesar et al., 2006; Husain, 2010; Oliveira et al., 2011). β-Galactosidases obtained from microorganisms have received particular attention due to their diverse properties, relative ease of preparation, and reduced cost. Most commercial β-D-Galactosidases are produced from *A. oryzae* and *Kluyveromyces lactis* (Nizamuddin et al., 2008; Cardelle-Cobas et al., 2011). β-Galactosidase from other fungi has been reported, such as *Kluyveromyces fragilis*, *Aspergillus niger*, and *Aspergillus carbonarius* (Bansal, et al., 2008; O'Connell et al., 2008; Chen, 2008). Numerous bacteria have been suggested as promising microbial β-Galactosidase producers, such as *Bacillus stearothermophilus*, *Bifidobacterium longum*, and *Arthrobacter* (Coker al., 2003; Chin-A, et al., 2006; Chen, 2008). The properties of β-Galactosidase from different microorganisms vary widely, particularly in effect of pH and temperature.

The yak is the only bovine species adapted to the cold and harsh conditions of the Hindu Kush-Himalayan region and the Qinghai-Tibetan plateau with its altitude range of 2000–5000m above sea level (Cui et al., 2016). In China, there are about 14 million yak distributed in Yunnan, Xizang, Qinghai, Gansu, Sichuan, and other provinces, accounting for 95% of the world's yak population. The Yak have been central to the development of the farming and pastoral communities of these areas (He et al., 2011). It has been widely reported that the nutrient content of yak milk, including fat, protein, and lactose, are higher than that of cow milk (Yan and Pan 2004). Therefore, yak milk is often preferred as "natural condensed milk" in Tibet (Chen et al., 2008). Kurut, the traditional, naturally-fermented yak milk, is one of the staple foods of the Tibetan nomadic people. Because Kurut is usually prepared with unpasteurized yak milk under open conditions, both endogenous and exogenous microorganisms are involved in the fermentation process. The microbial population inhabiting traditional fermented yak milk is very diverse (Sun et al., 2010). It is possible to find microorganisms producing β-Galactosidases in the traditional fermented yak milk.

In the present study, microorganisms producing β-Galactosidases were isolated from Yak Yoghourt samples collected from Gannan pasturing area of Gansu Province. In addition, the isolation and identification of the microorganism was carried out. On this basis, enzyme properties of the β-Galactosidase produced by *Enterobacter* sp. SYA2 were investigated.

2. MATERIAL AND METHODS

2.1. Collection of Samples

A total of 25 samples of traditional fermented Kurut were collected from different households in the Gannan Tibetan Autonomous Region, China. The samples were aseptically placed in sterile bottles, leveled, stored in an ice-box container and transported to the laboratory within 48 hours for the purpose of isolation of strains.

2.2. Isolation of Bacterium Producing β-Galactosidases

Screenings were carried out by the spread-plate methods. Kurut samples were suspended and serially diluted in sterile saline, and aliquots were spread onto the selective medium (2% lactose, 0.5% Bacto yeast extract, 1% Bacto peptone, 0.03% X-gal, and 2% agar). Plates were incubated for 24 hours at 37°C. The colonies that exhibited blue colors were selected, and streaked on additional new plates until homogenous colonies were observed. The purified strains were stored at 0–4°C.

2.3. Identification of Bacterium Producing β-Galactosidases

The detection of physiological and biochemical characteristics was performed based on Bergey's Manual of Determinative Bacteriology (Holt et al., 1994). Morphological properties were examined by using a microscope. The selected isolate was identified using 16S rDNA. Homology searches were performed against the sequences with the database using the BLAST program (NCBI). For phylogenetic analysis, a dataset containing GenBank 16S rRNA gene sequences were aligned using the CLUSTALX program (Thompson et al., 1997). To construct the phylogenetic tree, MEGA 4.0 software was used (Tamura et al., 2007).

2.4. Preparation of Cell Extracts

Cells grown in a shake flask containing the liquid medium (2% lactose, 0.5% Bacto yeast extract, 1% Bacto peptone) were collected and concentrated by centrifugation at 6,000×g for 20 min. The supernatant was discarded, and the cells were washed with one volume of 0.1 M potassium phosphate buffer (pH 6.5) and then centrifuged again. The pellets were suspended in one volume of the above buffer, broken by sonication, and centrifuged at 12,000×g for 20 min. The supernatant was obtained for β-Galactosidase analysis or purification.

2.5. Assay of β-Galactosidase Activity

β-Galactosidase activity was assayed in 100 mM potassium phosphate (pH 6.5) using the chromogenic substance o-nitrophenyl-β-D-galactopyranoside (ONPG) as substrate. The reaction mixture was incubated at 37°C for 10 min. The enzyme reaction was stopped by adding 0.5 M Na_2CO_3. The increase in absorbance at 420 nm due to the liberation of o-nitrophenol (ONP) was measured spectrophotometrically. One unit of β-Galactosidase activity was defined as the amount of enzyme that produced 1 μmol of ONP per minute.

2.6. Preparation of Crude Enzyme

The crude enzyme extract was precipitated with ammonium sulfate (30%–90% saturation). The precipitate obtained after centrifugation at 12,000g for 15 min was suspended in 50 mM sodium phosphate buffer (pH 7.0) and dialysed (7 kDa cutoff) overnight against several changes of distilled water to remove the salt.

2.7. Effects of Temperature and pH on Enzyme Activity and Stability

The optimum temperature for the activity of the crude enzyme was determined by assaying the β-Galactosidase activity at intervals of 5°C in the temperature range 30–60°C. The optimum pH for the activity of the enzyme was determined by assaying the β-Galactosidase activity in the pH range 4.0–9.0, by adjusting the pH of the substrate with 0.1 M HCl or 0.1 M NaOH as appropriate. The maximum β-Galactosidase activity obtained was taken to be 100%.

To determine the thermal stability, the purified enzyme was incubated at 5°C intervals in the temperature range 35–50°C, and the length of the incubation was varied from 0 to 60 min. After incubation, the residual β-Galactosidase activity was determined and the activity obtained with an incubation time of 0 min was taken to be 100%.

To determine the pH stability, the enzyme was dispersed (1:1, v/v) in the following 0.1 M buffer solutions: glycine–HCl (pH 4.0), citrate/phosphate (pH 4.5–5.5), sodium phosphate (pH6.0–8.5) and carbonate/bicarbonate (pH9.0), and kept at room temperature for 24 h. The residual β-Galactosidase activity was determined and the maximum activity obtained was taken to be 100%.

2.8. Effect of Metal Ions

The effect of metal ions Na^+, K^+, Mg^{2+}, Ca^{2+}, Fe^{2+}, Zn^{2+}, Mn^{2+} and Cu^{2+} on the β-Galactosidase activity was determined at metal ion concentrations of 1 mM, 5 mM and 10 mM. The β-Galactosidase was incubated at room temperature for 40 min with metal ions. The β-Galactosidase activity obtained without metal ions was taken to be 100%.

2.9. Statistical Analysis

All experiments were done in triplicate and the results are expressed as mean ± standard deviation. The data were analysed by one-way ANOVA using SPSS version 17.0 and the level of statistical significance was set at $P < 0.05$.

3. RESULTS AND DISCUSSION

3.1. Screening of Bacterium Producing β-Galactosidase

During screening process, a total of 21 isolates were obtained for β-Galactosidase production via X-gal hydrolysison plate. β-Galactosidase activity of the strains ranged from 1.11 to 6.23 U/mL(Table 1). Among them, five strains' β-Galactosidase activity was higher than 4.0 U/mL. An isolate SYA2 had highest enzyme activity (6.23 U/mL) and was selected for further study.

3.2. Identification of the Selected Strains

The strain SYA2was Gram-negative, aerobic, and non-sporing, and predominantly formed smooth-looking white colonies on beef-peptone medium after 24h of incubation (Figure 1, Figure 2). It was able to ferment glucose, sucrose, and fructose.

Table 1. Lactase activity of 21 strains

Strain number	Enzyme activity (U/mL)	Strain number	Enzyme activity (U/mL)	Strain number	Enzyme activity (U/mL)
SYA2	6.23±0.015^a	TB9	3.28±0.013^h	TD9	2.43±0.035^m
TB2	4.81±0.017^b	TC5	3.14±0.018^i	TD17	2.40±0.026^m
TG25	4.41±0.020^c	TB6	2.86±0.018^j	SB2	2.34±0.0088^n
TA8	4.25±0.014^d	SB6	2.71±0.006^k	TD25	2.27±0.013^o
TE1	4.04±0.026^e	TA1	2.69±0.022^k	SB3	2.13±0.021^p
TF2	3.82±0.013^f	TA10	2.48±0.023^l	SD3	1.32±0.021^q
SG2	3.66±0.011^g	SME1	2.48±0.026^l	TF12	1.11±0.031^r

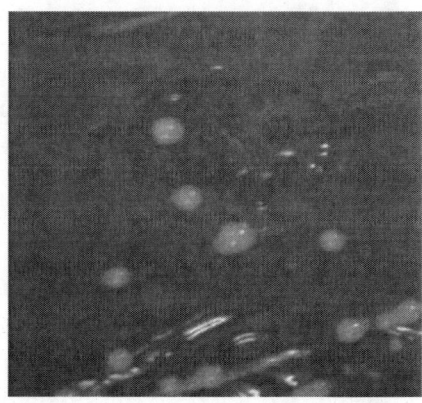

Figure 1. Colony morphology of strain SYA2 on beef-peptone medium.

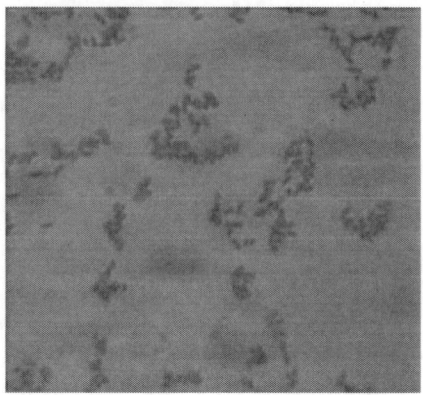

Figure 2. Gram stain of strain SYA2.

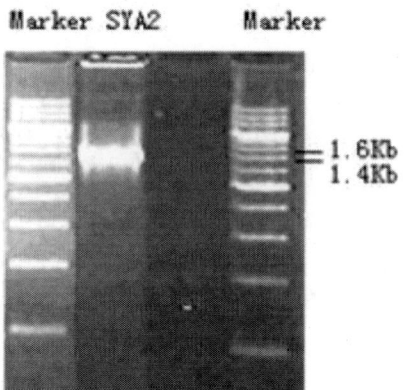

Figure 3. Agarose gel electrophoresis pattern of PCR amplification product of 16s rDNA.

The strain was identified using 16S rDNA analysis, agarose gel electrophoresis of PCR product amplified from strain SYA2 was showed in Figure 3. The 16S rDNA gene sequences of SYA2 (1414 bp) were compared to all sequences in GenBank and its GenBank accession number was JN231311. The nucleotide sequence showed the closest match to that of *Enterobacter* sp. dc6 with a homology of 99%. For phylogenetic analysis, the phylogenetic tree was constructed using neighbor-joining (Figure 4). Confidence in the tree topology was determined by bootstrap analysis using 1,000 resampling of the sequences. Therefore, the isolate

SYA2 could be identified as *Enterobacter* sp. *Enterobacter* sp. SYA2 has been deposited with the China General Microbiological Culture Collection Center; CGMCC NO.5251. The microorganism was maintained in the laboratory at 37°C on LB agar slants (1.0% (w/v) peptone, 1.0% (w/v) beef extract, 0.5% (w/v) NaCl and 2.0% (w/v) agar, with pH 7.2).

3.3. Time Course of β-Galactosidase Production by *Enterobacter* sp. SYA2

As Figure 5 showed, β-Galactosidase activity increased linearly with increasing incubation time till 24h, further incubation showed reduction in enzyme production. The reduction in enzyme production after optimum period was probably due to depletion of nutrients available to microorganisms. Several researchers reported that the optimum fermentation periods for the production of β-Galactosidase were between 24-48 h. Although it might be insignificant to make such a comparison when the fermentation conditions were different, it indicated that the fast production of β-Galactosidase by *Enterobacter* sp. SYA2 reported here places an advantage for industrial purpose, as compared to other sources of β-Galactosidase.

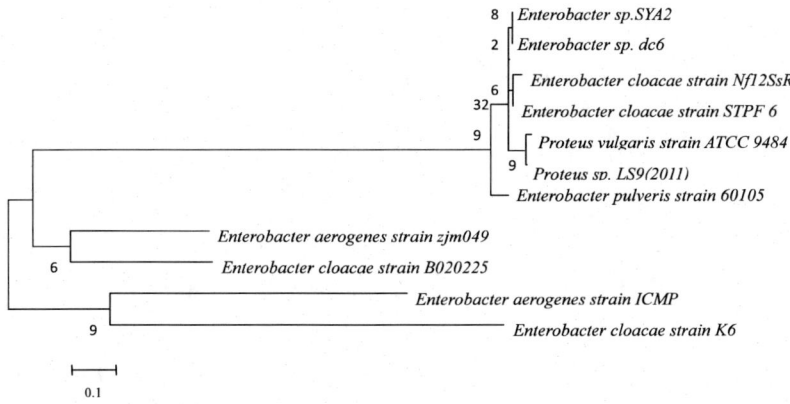

Figure 4. Phylogenetic tree for SYA2 and related strains.

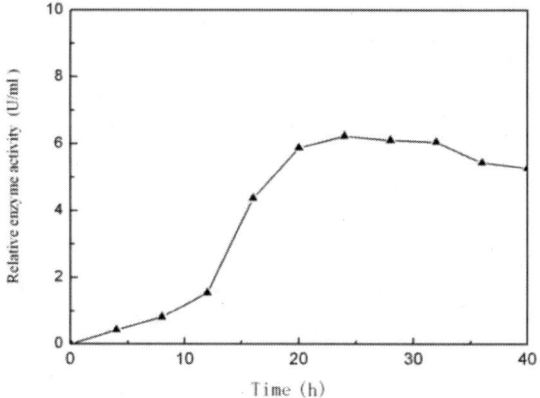

Figure 5. The time course of β-Galactosidase production of *Enterobacter* sp. SYA2.

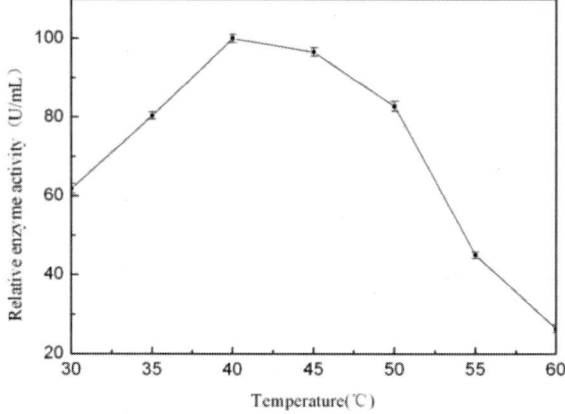

Figure 6. Effect of temperature on the β-Galactosidase activity. Error bars represent the standard deviation of triplicate experiments.

3.4. Effects of Temperature and pH on Enzyme Activity

The β-Galactosidase activity increased with increased temperature in the temperature range 30–40°C and the optimum temperature for the purified enzyme was 40°C (Figure 6). Different enzymes have different optimum temperatures, mainly depending on the enzymes' structure. This result was in accord with the β-Galactosidase from *Escherichia coli K12*

(Gary et al., 1965), but different from *Lactobacillus crispatus* (Him and Rajagopal, 2000), which has an optimum temperature in the range 45°C. This substantial difference in optimum temperature between the microbial β-Galactosidase suggests strongly that they are suitable for use under different conditions.

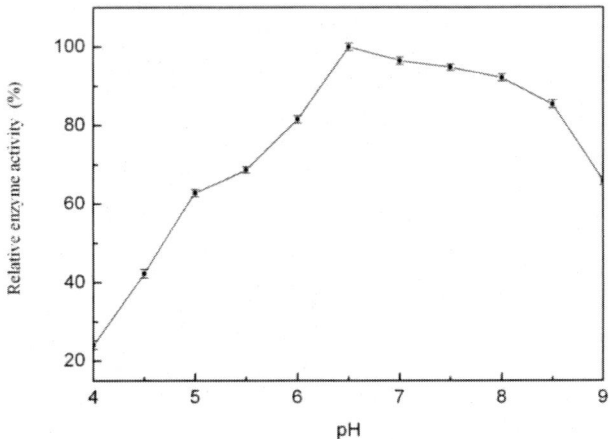

Figure 7. Effect of pH on the β-Galactosidase activity. Error bars represent the standard deviation of triplicate experiments.

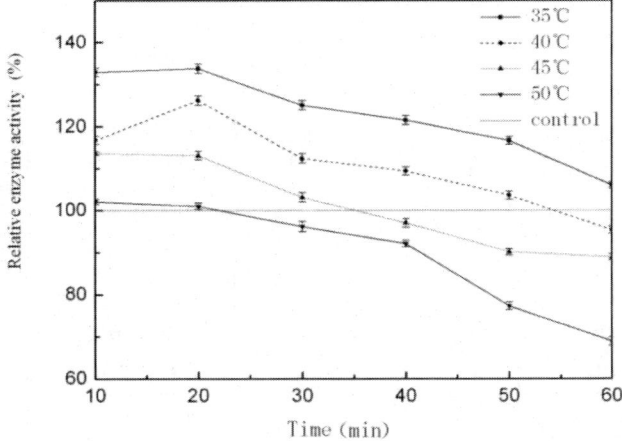

Figure 8. Heat stability of the β-Galactosidase. Error bars represent the standard deviation of triplicate experiments.

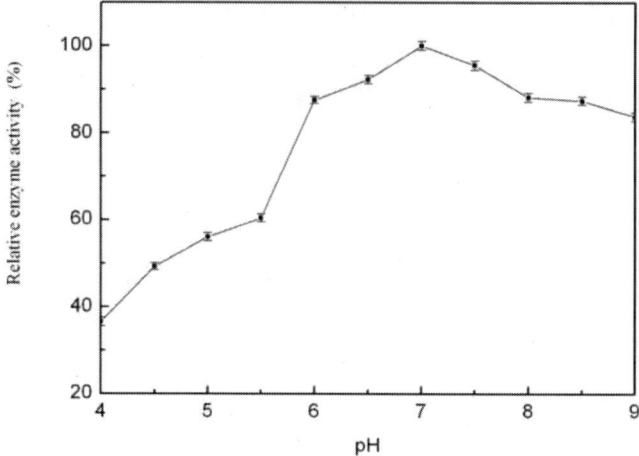

Figure 9. pH stability of the β-Galactosidase. Error bars represent the standard deviation of triplicate experiments.

The crude enzyme from *Enterobacter sp.* SYA2 had a higher level of β-Galactosidase activity in the neutral range (6.5-8.5). The maximum β-Galactosidase activity was at pH 6.5, and the activity decreased with increasing pH (Figure 7). This result was similar to what is reported for the β-Galactosidase enzymes from *Lactobacillus crispatus* (Him and Rajagopal, 2000), but different from *Aspergillus candidus* (Connell and Walsh, 2008), which has an optimum pH at 4.5.

3.5. Effects of Temperature and pH on Enzyme Stability

The heat stability of the enzyme is shown in Figure 8. It was fully active after 60 min of incubation at 35°C but inactivated after 60 min at 50°C. Compared with the enzyme from *Bacillus* sp. MTCC-864 (Patil et al., 2011), which completely lost its activity after 30 min at 50°C, and *Ranhella aquatilis* KNOUC601 (Nam and Ahn, 2011), which lost their activity after 10 min at 37°C, the enzyme has a higher level of thermostability. The higher level of thermostability of the *Enterobacter* sp. SYA2enzyme made it easy to be used.

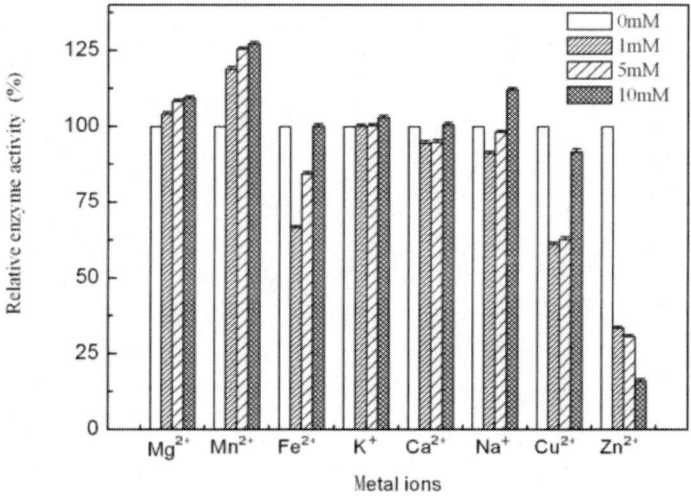

Figure 10. Effect of metal ions on the β-Galactosidase activity. Error bars represent the standard deviation of triplicate experiments.

The stability of the enzyme at different pH values is shown in Figure 9. The enzyme was stable in a relatively wide range of pH 6.5–9.0, with maximum stability at pH 7.0. Outside either end of this range, the activity of the enzyme decreased drastically, which showed that the *Enterobacter* sp. SYA2enzyme was stable in the neutral range. The low pH-sensitivity of the *Enterobacter* sp. SYA2enzyme is useful for the food industry.

3.6. Effect of Metal Ions

Figure 10 showed that different metal ions had different effects on β-Galactosidase activity. Cu^{2+} and Fe^{2+} had only a slight inhibitory effect, whereas Zn^{2+} inhibited the milk-clotting activity significantly. In contrast, Mn^{2+} and Mg^{2+} had a significant stimulatory effect on milk-clotting activity, whereas 10 mM Na^+ promoted the activity slightly, but had a slight inhibitory effect when present at concentrations of only 1 mM or 5 mM. K^+ and Ca^{2+} had no effect when present at different concentrations.

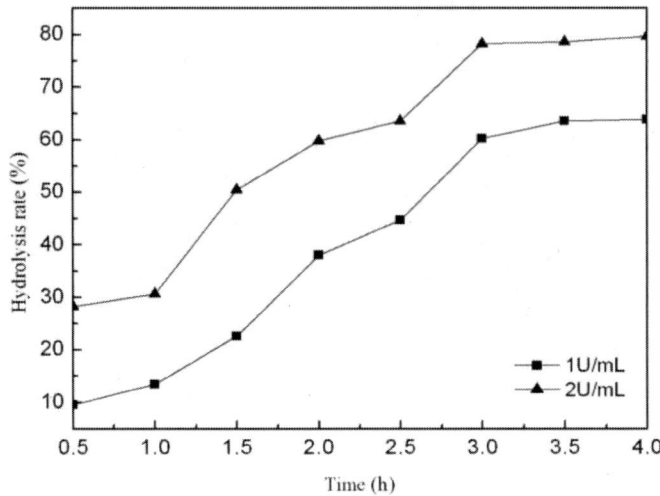

Figure 11. Hydrolysis of lactose in milk at different β-Galactosidase concentration.

3.7. Hydrolysis of Lactose in Milk

Experiments on lactose hydrolysis in milk were performed by β-Galactosidase from *Enterobacter* sp. SYA2. After 1hour of incubation at 40°C, 13.41% of milk lactose was hydrolyzed by β-Galactosidase at enzyme concentration of 1 Um/L, which increased to 63.78% in 4 h. More than 30.66% lactose hydrolysis was achieved in 1 hour at enzyme concentration of 2 Um/L, which increased to 79.62% in 4 h. These results probably indicate that SYA2 has a more notable effect on lactose hydrolysis in milk and whey in dairy industry.

CONCLUSION

In this study, strains with β-Galactosidase activity were isolated from 25 Yak Yoghourt samples collected from Gannan pasturing area of Gansu Province. An isolate with good β-Galactosidase activity identified as *Enterobacter* sp. The β-Galactosidase produced by *Enterobacter sp.* SYA2 was purified by ammonium sulfate precipitation. The optimum temperature, thermostability, optimum pH and pH stability of the β-Galactosidase, as well as the effects of metal ions on enzymatic activity, were investigated. These results suggest that *Enterobacter* sp. SYA2 is a possible commercial source of β-Galactosidase for lactose hydrolysis in milk.

ACKNOWLEDGMENTS

This work was supported by National Natural Science Fund of China (31560442, 31760466,31460425),Special funds for discipline construction of GAU (GAU-XKJS-2018-247), Program for Fu Xi Talents in GAU (FXYC20130110), the and the Enterprise research transformation and industrialization project (No.2018-SF-C29) for financial support.

REFERENCES

Bansal, S., Oberoi, H.S., Dhillon, G.S. & Patil, R.T. (2008). Production of β-galactosidase by *Kluyveromyces marxianus* MTCC 1388 using whey and effect of four different methods of enzyme extraction on β-galactosidase activity. *Indian Journal of Microbiology*, 48(3),337–341.

Cardelle-Cobas, A., Corzo, N., Martinez-Villaluenga, C., Olano, A. & Villamiel, M. (2011). Effect of reaction conditions on lactulose-derived trisaccharides obtained by transgalactosylation with β-galactosidase of *Kluyveromyces lactis*. *European Food Research and Technology*, 233(1), 89–94.

Chen, W., Chen, H., Xia, Y., Zhao, J., Tian, F. &Zhang, H. (2008). Production, purification, and characterization of a potential thermostable galactosidase for milk lactose hydrolysis from *Bacillus stearothermophilus*. *Journal of Dairy Science*, 91(5), 1751–1758.

Chen, Z. L., Cheng, C., Ma, K., Liu, G. Q., Li, H. & Li, H. (2008). Isolation and identification of lactic acid bacteria from fermented yak milk products in Tibet area. *Food Science*, 29(12), 408-412.

Chin-A, H., Roch-Chui, Y. & Cheng-Chun, C. (2006). Purification and characterization of a sodium-stimulated β-galactosidase from *Bifidobacterium longum* CCRC 15708. *World Journal of Microbiology and Biotechnology*, 22(4), 355–361.

Coker, J.A., Sheridan P. P., Loveland Curtze, J., Gutshall, K.R., Auman, A.J. & Brenchley, J.E. (2003). Biochemical characterization of a β-galactosidase with a low temperature optimally obtained from an Antarctic Arthrobacter isolate. *Journal of Bacteriology*, 185(18), 5473–5482.

Connell, S. O. &Walsh, G. (2008). Application Relevant Studies of Fungal β-galactosidases with Potential Application in the Alleviation of Lactose Intolerance. *Application Biochemistry Biotechnology*, 149(2),129-138.

Cui, G. X.,Yuan, F., Degen, A. A., Liu, S. M., Zhou, J. W., Shang, Z. H., Ding, L. M., Mi, J. D., Wei, X. H. & Long, R. J. (2016). Composition of the milk of yaks raised at different altitudes on the Qinghai–Tibetan Plateau. *International Dairy Journal*, 59, 29–35.

Gary, R., Edward, S. & Christian, B. (1965). Purification, composition and molecular weight of the β-galactosidase of *Escherichia coli* K12. *Biological Chemistry*, 240(6), 2468-2477.

He, S. H., Ma, Y., Wang, J. Q., Li, Q. M., Yang, X., Tang, S. H. & Li, H. M. (2011). Milk fat chemical composition of yak breeds in China. *Journal of Food Composition and Analysis*, 24(2), 223-230.

Him, J. W. & Rajagopal, S. N. (2000). Isolation and characterization of β-galactosidase from *Lactobacillus crispatus*. *Folia Microbiologica*, 45(1), 29-34.

Holt, J. G., Krieg, N. R., Sneath, P. H. A., Staley, J. T., Williams, S. T. (1994). Irregular, nonsporing Gram-positive rods. In: *Bergey's manual of determinative bacteriology*, 9th Edn. Williams and Wilklins, Baltimore, 274-553.

Husain, Q. (2010). β-Galactosidases and their potential applications: a review. *Critical Reviews in Biotechnology, 30*(1),41-62.

Nam, E. S. & Ahn, J. K. (2011). Isolation and Characterization of Cold-adapted bacteria Producing Lactose hydrolyzing enzyme isolated from soils of Nome area in Alaska. *International Research Journal of Microbiology, 2*(9), 348-355.

Nizamuddin, S., Sridevi, A. & Narasimha, G. (2008). Production of β-D-galactosidase by *Aspergillus oryzae* in solid-state fermentation. *African Journal of Biotechnology, 7*(8), 1096–1100.

O'Connell, S. & Walsh, G. (2008). Application relevant studies of fungal β-galactosidases with potential application in the alleviation of lactose intolerance. *Applied Biochemistry and Biotechnology, 149*(2), 129–138.

O'Connell, S. & Walsh, G. (2010). A novel acid-stable, acid-active β-galactosidase potentially suited to the alleviation of lactose intolerance. *Applied Microbiology and Biotechnology, 86*(2), 517–524.

Oliveira, C., Guimaraes, P.M.R. & Domingues, L. (2011). Recombinant microbial systems for improved β-galactosidase production and biotechnological applications. *Biotechnology Advances, 29*(6), 600-609.

Panesar, P.S., Panesar, R., Singh, R.S., Kennedy, J.F. & Kumar, H. (2006). Microbial production, immobilization and applications of β-D-galactosidase. *Journal of Chemical Technology & Biotechnology, 81*(4), 530-543.

Patil, M. M., Mallesha, K. V. R. & Bawa, A. S. (2011). Characterization of partially purified β-galactosidase from *Bacillus* sp. MTCC-864. *Recent Research in Science and Technology, 3*(9), 84-87.

Sun, Z., Liu, W., Gao, W., Yang, M., Zhang, J., Wu, L., Wang, J., Menghe, B., Sun, T. & Zhang, H. (2010). Identification and characterization of the dominant lactic acid bacteria from kurut: The naturally fermented yak milk in Qinghai, China. *The Journal of General and Applied Microbiology, 56*(1), 1–10.

Tamura, K., Dudley, J., Nei, M. & Kumar, S. (2007). MEGA4: molecular evolutionary genetics analysis (MEGA) software version 4.0. *Molecular Biology and Evolution*, 24(8), 1596-1599.

Thompson, J. D., Gibson, T. J., Plewniak, F., Jeanmougin, F. & Higgins, D. G. (1997). The CLUSTALX windows interface: flexible strategies for multiple sequence alignment aided by quality analysis tools. *Nucleic Acids Research*, 25(24), 4876-4882.

Yan, P. & Pan, H. P. (2004). Development and utilization of yak milk. *Agriculture Products Development*, 7, 10-12.

INDEX

A

active site, 11, 12, 18, 19, 35, 75, 76, 79, 82, 92, 121, 122
aeration, viii, 2, 4, 20, 22, 25, 26, 42
agitation, viii, 2, 4, 20, 21, 22, 25, 26, 42
alkylglycosides, 95
animal β-Galactosidases, 69
Arabidopsis thaliana, 71, 103
arabinogalactan, 104
Aspergillus oryzae (*A. oryzae*), ix, 8, 10, 16, 21, 38, 39, 41, 42, 45, 46, 48, 57, 61, 62, 66, 72, 75, 77, 80, 84, 86, 88, 89, 91, 93, 94, 95, 96, 97, 98, 99, 100, 105, 106, 107, 108, 110, 112, 113, 114, 115, 125, 135, 142, 147, 148, 149, 150, 151, 152, 153, 165, 168, 183

B

B. circulans, 85, 86, 88, 93, 94, 95, 98, 125
bacteria, viii, ix, 2, 6, 10, 11, 12, 14, 20, 26, 27, 28, 30, 31, 33, 40, 65, 68, 69, 72, 73, 74, 75, 125, 143, 163, 165, 168, 182, 183
biomolecules, 4, 15, 20, 27, 28, 30
bioprocess parameters, v, 14, 21
biosensors, 100
biosynthesis, 10, 98
biotechnological applications, 104, 154, 183
biotechnology, 2, 41, 42, 43, 44, 45, 48, 49, 51, 52, 53, 54, 56, 58, 59, 75, 117, 128, 140, 146, 148, 149, 150, 151, 152, 153, 154, 155, 156, 158, 159, 160, 162, 163, 165, 166, 182, 183

C

carbohydrate metabolism, 143
carbon, viii, 2, 4, 15, 16, 26, 72, 82, 85, 156
catalysis, viii, 2, 18, 76, 91, 158
catalytic activity, 35, 36, 76, 106
cation, 76, 135, 146, 157
cell disruption, 4, 10, 29, 30, 31, 32, 33, 34, 43, 48, 56
cellulose, 36, 38, 39, 106
chitosan, 37, 38, 58, 59, 60, 152
chromatography, 28, 29, 131, 133, 134, 135, 142, 157, 158
cofactors, viii, 2, 18, 80
colon, 69, 143
colorectal cancer, 160

composition, 17, 18, 23, 30, 34, 69, 93, 106, 118, 122, 123, 142, 161, 182
compounds, 18, 67, 93, 94, 96, 97, 99, 109, 118, 119, 133, 134, 142, 159
consumption, ix, 3, 23, 66, 103, 119, 132, 135, 136, 137, 139
consumption rates, 23
contamination, 34, 35, 91, 138
crystal structure, 77, 105, 147
crystalline, 5
crystallization, viii, ix, 1, 3, 13, 39, 66, 90, 130
culture medium, viii, 2, 4, 14, 15, 16, 18, 20, 21, 22, 23, 24, 25, 26, 27, 28, 52

enzyme immobilization, 39, 92, 126, 151, 152, 164
epithelial cells, 6
equilibrium, 5, 83, 84
Escherichia coli (*E. coli*), ix, 9, 11, 12, 35, 49, 50, 51, 66, 68, 74, 75, 76, 79, 80, 85, 86, 87, 88, 89, 95, 97, 98, 101, 104, 105, 106, 107, 108, 110, 115, 147, 157, 176, 182
ethanol, 3, 32, 33, 86, 89, 111, 131, 133, 136, 155, 156
ethylene, 99, 102, 104
ethylene glycol, 99
eukaryotic, 68

D

dairy industry, 10, 90, 111, 180
downstream, v, vii, viii, 1, 2, 4, 27, 28, 30, 44, 45, 47, 54, 55, 57, 137, 139
drugs, x, 3, 6, 12, 66

E

E. coli, 12, 35, 75, 79, 85, 86, 87, 88, 89, 95, 97, 98, 101, 105
Enterobacter sp. SYA2, xi, 168, 169, 175, 176, 178, 179, 180
enzymatic activity, viii, xi, 2, 5, 7, 11, 16, 17, 18, 19, 20, 26, 29, 39, 76, 168, 181
enzyme(s), vii, 1, 2, 3, 4, 5, 6, 7, 10, 11, 12, 13, 15, 16, 17, 18, 19, 26, 27, 29, 33, 34, 35, 36, 37, 38, 39, 40, 41, 42, 44, 50, 51, 56, 58, 60, 67, 68, 70, 72, 75, 79, 80, 81, 82, 83, 84, 85, 87, 88, 89, 90, 91, 92, 93, 94, 95, 99, 100, 103, 106, 107, 108, 110, 111, 112, 113, 114, 115, 120, 121, 122, 123, 124, 125, 126, 127, 128, 129, 130, 141, 142, 148, 149, 151, 152, 153, 154, 158, 164, 169, 171, 172, 173, 175, 176, 178, 179, 180, 181, 183

F

fermentation, viii, 2, 4, 14, 15, 20, 22, 23, 24, 26, 111, 129, 136, 137, 143, 146, 158, 169, 175, 183
food, viii, x, 1, 3, 4, 12, 34, 35, 40, 66, 75, 90, 94, 95, 104, 111, 112, 118, 128, 129, 136, 143, 144, 145, 156, 157, 158, 159, 161, 162, 163, 179
food additives, x, 66
Food and Drug Administration, 4
food industry, viii, 1, 12, 35, 40, 94, 129, 136, 161, 179
food processing industry, 3, 4, 40
food production, 159
food products, 3
functional food, x, 92, 118, 119
fungi, viii, ix, 2, 4, 7, 8, 10, 12, 13, 20, 40, 65, 72, 75, 80, 168

G

galactooligosaccharides, x, 3, 13, 45, 53, 58, 66, 92, 100, 111, 145, 147, 149, 150, 151, 155, 156, 157, 158

galactosylation, x, 66, 86, 96, 97, 98, 108, 114
galectin inhibitors, x, 66, 99
GH families, 67
glucose, vii, viii, ix, 1, 5, 6, 12, 15, 16, 17, 19, 22, 25, 26, 30, 66, 69, 80, 81, 83, 85, 88, 90, 92, 118, 120, 121, 122, 124, 125, 126, 129, 130, 136, 142, 150, 168, 173
glucose oxidase, 129
glucoside, 85
glutamic acid, 12, 82, 121, 122
glycans, 78, 100
glycerol, 88, 89, 96, 99, 110
glycine, 172
glycol, 100, 113
glycoproteins, 68, 70, 100, 102, 115
glycoside, ix, 65, 67, 68, 82, 108, 142, 146
glycosylation, 77, 115

H

human health, x, 118, 143
human milk, 119, 136, 158
human skin, 161
hydrogen, 22, 37, 69, 78, 103, 122
hydrogen bonds, 122
hydrolysis, vii, viii, ix, xi, 1, 4, 6, 10, 11, 12, 14, 15, 36, 39, 66, 67, 68, 69, 70, 80, 81, 83, 84, 87, 88, 90, 91, 93, 106, 107, 110, 112, 113, 119, 120, 121, 122, 124, 127, 130, 136, 139, 142, 146, 148, 149, 152, 154, 155, 158, 168, 172, 180, 181
hydrolyzed collagen, 17
hydrophilicity, x, 66, 96
hydrophobicity, 28
hydroquinone, 86, 96, 109

I

identification, vi, 58, 113, 150, 167, 169, 170, 173, 182, 183

immobilization, vii, ix, 2, 4, 13, 34, 35, 36, 37, 38, 39, 40, 41, 43, 47, 48, 50, 53, 54, 57, 59, 60, 63, 91, 92, 101, 106, 111, 115, 126, 149, 150, 151, 152, 153, 162, 163, 164, 165, 166, 183
immobilized enzymes, ix, 2, 13, 37, 38, 39, 92, 100
immune system, 119, 143, 144
immunoglobulin, 78
in vitro, 6, 146, 156, 160
in vivo, 6, 98, 114, 146, 157
inflammatory bowel disease, 6, 143

K

kinetic parameters, 23, 38
Kluyveromyces lactis, ix, 10, 22, 44, 47, 49, 51, 53, 55, 58, 62, 66, 68, 72, 78, 79, 80, 105, 106, 108, 109, 110, 113, 136, 147, 152, 153, 168, 181

L

lactase, vii, 1, 3, 6, 12, 13, 68, 69, 72, 103, 109, 112, 114, 168
lactic acid, 11, 26, 30, 31, 33, 40, 86, 125, 165, 182, 183
Lactobacillus, 9, 11, 16, 17, 21, 25, 30, 33, 41, 42, 44, 49, 51, 52, 55, 57, 58, 61, 68, 73, 74, 80, 92, 03, 106, 107, 109, 150, 151, 155, 164, 177, 178, 182
lactose, vii, viii, ix, xi, 1, 3, 4, 5, 6, 10, 11, 12, 13, 14, 15, 16, 17, 21, 22, 23, 24, 25, 36, 38, 39, 42, 44, 45, 46, 47, 48, 49, 50, 51, 52, 53, 54, 55, 56, 57, 58, 59, 60, 61, 62, 66, 68, 69, 72, 80, 81, 84, 85, 86, 88, 90, 91, 92, 93, 94, 95, 96, 97, 98, 99, 103, 106, 107, 108, 110, 111, 112, 113, 119, 120, 121, 122, 123, 124, 125, 126, 127, 129, 130, 131, 133, 136, 137, 139, 141, 142, 145, 146, 147, 148, 149, 150,

151, 152, 153, 154, 155, 157, 158, 168, 169, 170, 180, 181, 182, 183
lactose hydrolysis, ix, xi, 4, 10, 12, 14, 15, 36, 39, 42, 45, 48, 50, 52, 57, 58, 59, 61, 66, 68, 80, 81, 88, 90, 93, 106, 112, 119, 120, 124, 148, 149, 152, 154, 158, 168, 180, 181
lactose intolerance, 3, 5, 13, 49, 51, 53, 55, 103, 182, 183
lactose-free, 3, 10, 12

M

magnesium, viii, 2, 7, 17, 19, 22, 79
metabolites, viii, 2, 23, 24, 27
metal ions, viii, xi, 2, 19, 28, 168, 172, 179, 181
Mg^{2+}, xi, 19, 80, 168, 172, 179
microbial β-galactosidase, v, 1, 7
microorganism(s), viii, ix, 2, 7, 8, 9, 11, 13, 18, 20, 21, 22, 23, 24, 25, 26, 31, 32, 35, 66, 80, 137, 169, 175
molecular weight, 92, 106, 132, 133, 138, 182
monosaccharide, 16, 118, 122, 129, 158

N

Na^+, xi, 19, 76, 80, 134, 168, 172, 179
NaCl, 32, 175
nitrogen, viii, 2, 4, 14, 15, 16, 17, 18, 24, 47, 90
nucleophilicity, 86
nucleotide sequence, 174
nucleotide sequencing, 105
nucleotides, 120

O

oligosaccharide, 78, 107, 108, 112, 134, 148, 149, 151, 153, 154, 155, 158, 160, 165
organic solvents, 28, 83, 109, 113, 129
oxygen, viii, 2, 14, 20, 23, 82, 91

P

pH, viii, ix, xi, 2, 4, 7, 8, 9, 10, 13, 15, 16, 17, 20, 21, 22, 23, 25, 28, 29, 37, 38, 66, 80, 90, 91, 106, 127, 128, 141, 143, 150, 162, 168, 170, 171, 172, 175, 176, 177, 178, 179, 181
plants, vii, viii, ix, 1, 3, 6, 7, 15, 40, 65, 68, 70, 71, 103, 104, 168
plants β-galactosidases, 70
polymerization, 118, 123, 125, 127, 132, 134, 142
polymers, ix, 30, 66, 100, 149
polypeptide, 3, 75, 77
polysaccharides, 70, 92, 102
polyvinyl alcohol, 112, 152
potassium, viii, 2, 17, 19, 21, 22, 170, 171
proteins, 14, 28, 30, 37, 67, 72, 91, 99, 101, 111, 141
purification, vii, viii, xi, 2, 11, 27, 28, 29, 30, 40, 74, 75, 91, 100, 105, 107, 118, 119, 130, 131, 132, 133, 135, 136, 137, 139, 141, 146, 149, 155, 156, 157, 158, 159, 160, 170, 181

R

reaction rate, 19, 127
reaction temperature, 127
reaction time, 83, 84, 122
reactions, 23, 81, 83, 92, 107, 110, 120, 121, 122, 125, 126, 129, 130, 155, 162, 163

Index

residues, 7, 12, 15, 17, 30, 67, 68, 70, 77, 78, 79, 82, 101, 104, 121

S

screening, vi, 41, 167, 168, 172
small intestine, 6, 69
sodium, 16, 17, 19, 22, 33, 79, 97, 171, 172, 182
sodium dodecyl sulfate, 33
solubility, viii, 2, 83, 96, 97, 123, 127, 129, 131
stability, ix, xi, 2, 3, 7, 11, 13, 29, 34, 35, 36, 37, 38, 39, 66, 95, 118, 119, 126, 127, 129, 149, 168, 172, 177, 178, 179, 181
stabilization, 35, 39, 152
structure, x, 18, 34, 66, 75, 76, 77, 78, 79, 84, 99, 100, 105, 109, 118, 123, 124, 125, 128, 176
symptoms, 3, 6, 13, 69
synthesis, ix, 11, 20, 38, 66, 82, 83, 84, 85, 86, 87, 88, 93, 94, 95, 97, 98, 99, 107, 108, 109, 110, 112, 113, 114, 119, 120, 122, 123, 125, 126, 127, 129, 137, 138, 143, 146, 149, 150, 151, 153, 154, 155, 158, 163, 164, 165

T

temperature, viii, ix, xi, 2, 4, 8, 9, 15, 20, 21, 22, 23, 24, 32, 37, 38, 39, 61, 66, 91, 106, 127, 138, 141, 149, 150, 154, 160, 168, 171, 172, 176, 178, 181, 182

thermal properties, 37
thermal stability, ix, 37, 66, 171
thermodynamic equilibrium, 82
thermogravimetric technique, 24
thermostability, xi, 168, 178, 181
transgalactosylation, v, x, 38, 51, 58, 61, 66, 82, 83, 84, 86, 88, 89, 92, 97, 98, 99, 108, 109, 110, 112, 113, 114, 117, 118, 120, 121, 122, 123, 124, 126, 127, 128, 150, 151, 154, 162, 163, 168, 181

U

urea, 18

V

viscosity, 28, 138, 143
vitamins, 4, 15

Y

yak yoghourt, vi, vii, xi, 167, 169, 180
yeast, ix, 4, 10, 12, 16, 17, 20, 21, 22, 25, 26, 30, 31, 33, 65, 72, 78, 80, 111, 136, 137, 158, 164, 170

Z

zirconia, 33

Related Nova Publications

CALMODULIN: STRUCTURE, MECHANISMS AND FUNCTIONS

EDITOR: Vahid Ohme

SERIES: Cell Biology Research Progress

BOOK DESCRIPTION: In *Calmodulin: Structure, Mechanisms and Functions*, the authors consider small and poorly-studied groups of plant calcium-dependent protein kinases that directly interact with calmodulin molecules.

SOFTCOVER ISBN: 978-1-53614-948-7
RETAIL PRICE: $82

MYOSIN: BIOSYNTHESIS, CLASSES AND FUNCTION

EDITOR: David Broadbent

SERIES: Cell Biology Research Progress

BOOK DESCRIPTION: *Myosin: Biosynthesis, Classes and Function* opens with a discussion on class I myosins, the most varied members of the myosin superfamily and a remarkable group of molecular motor proteins that move actin filaments and produce force.

SOFTCOVER ISBN: 978-1-53613-817-7
RETAIL PRICE: $95

To see complete list of Nova publications, please visit our website at www.novapublishers.com